Preface:

In terms of physics, in this manuscript we focus primarily on gravitational physics, with emphasis on applied physics, quantum theory (quantum mechanics and quantum field theory), mechanics and analytical mechanics. The mathematics consists of calculus, linear algebra, group theory, and tensor calculus.

Contents

Table of Contents:

Contents

Contents

modern formulation

Contents

Contents

Contents

Contents

Book 1:

Modern Formulation of the G.R.F.F.

(Gravitational Radiation Flux Force)

1

Introduction to an abstract derivation of a mechanical theory of gravity

Introduction

Generally speaking, there are four forms associated with the notion of an existing gravitational field. The first one involves the attraction between either two bodies of equal mass or one or more bodies of greatly differing mass in the macroscopic sense. This generally relates the gravitational behavior between a star and an orbiting planet. The second involves the gravitational attraction between a planetary body and any moon or satellite. This is the well-known three-body problem. The third is what is known as surface gravity which involves a singular body in relation to its angular rotation. The fourth we deem as a quantum gravitational field defined in the microscopic sense. We define this in relation to modern quantum field theory.

In the classical sense gravity can be defined as the physical force or field of force in which the planets are fixed in their orbits, stars and galaxies are grouped together, and the law of attraction holds between any two bodies of differing mass. That is to say the inert mass of a massive stellar object obtains a degree of attraction defined by the inert gravitational field that is associated with its mass and internal density. The first comprehensive theory of gravity was of course Isaac Newton's theory of gravity. This exhibited gravity via some intangible external force. But this existing force could not explain the cause and reason of the nearly uniform orbital *motion* of orbiting planets and existing stellar debris. The modern post-Newtonian metric theories of gravity exhibited the same difficulties, especially with the tides or current forces due to the pull of the moon. The difficulties with magnetic theories of gravity correspond to the following key examples. The first reason was that the planet Mars had no existing or measurable magnetic field and also had an orbiting moon. The second reason was that the Sun's magnetic field did not extend beyond the *heliopause* where in lay the lengthy room for orbiting bodies in the Oort cloud. For planets such as the earth, orbiting a center of gravity, the sun, they form orbital ellipses through their central motion in which they rotate with a consistent angular velocity on a given axis as they orbit at a constant velocity. The center of gravity of our solar system, our chosen stellar coordinate system, or system of reference, is the sun which includes all orbiting planets.

The diameter of the orbits of the planets can be computed in proportion to their orbiting angular velocity and their mass. The diameter stays constant, a fixed value, while the relative position of the perihelion of each planet varies with position corresponding each orbital, as defined by the distance separating the planets and the Sun. We know from Newton's gravitational law that two given bodies, regardless of mass, will attract each other with a defined force that is proportional to the product oftheir masses and inversely proportional to the square of the distance separating them. Just as there is a gravitational field uniformly surrounding the Sun and any other given star, whose magnitude and length as determined by chosen vector fields is deemed as proportional to the mass of the Sun or star. It is within these parameters that the linear inertial motion of the bodies situated within free space changes coarse by the effect of the field acting on the given body causing it to alter its original coarse of motion. The centrifugal force, central forces in nature, acts on the orbiting bodies at each given instant so that the motion remains uniformly elliptical.

We need to find a theory of gravity that complies with Newton's laws of motion (in particular the third law of motion) and Kepler's laws of planetary motion. The Cavendish experiment corresponds to the Newton's third law of motion. If the larger spheres were not fixed relative to the surface of the Earth they would exhibit no torsion to the suspending string for the smaller spheres suspended within close range to the fixed larger spheres (metallic spheres). First, we know in respect to the notion of the blackbody that all objects absorb and release energy usually in the form of infrared radiation. Massive stellar objects such as the Sun release energy in the form of a continuous spectrum radiation and solar wind. Now this radiation is released and absorbed within a 360-degree cone. And along a two-dimensional phase space the wave separates or joins in correlation to the inverse-square-law of universal gravitation. That is to say, at a distance N, from the Sun, the wave is separated within the phase space a distance D. Now at this distance N and D, it will exert a force of radiation pressure equivalent to the intensity I. Than in correlation to the inverse-square-law at a distance $2N$ we will have for the corresponding variables, $2D$ and $\frac{1}{2}I$. We also know that cosmic rays, small enough to pass through the *heliopause,* move toward the center of the Sun. We know this because the cosmic rays are found to land on surface of the Earth perpendicular to the surface throughout its orbital motion. And at a given distance from the Sun (within range of its gravitational field) there are equivalent pressures place from both sides perpendiculars to the Sun at any given point. And we know when two equivalent pressure are placed on a spherical object (pointed toward and away from the Sun) it will remain stationary. But if there is a minor angle variation with respect to the surface area in which the radiation strikes, as determined by the arbitrary factor pi, it will "slip" away and begin to orbit the Sun. Because of the movement of the revolving object, the center line of force, the radiation, is deviated slightly from the exact center of the spherical revolving object still pointed directly towards the center of gravity of the heliocentric system, the Sun, by a factor of pi. For each object throughout its orbital motion, on both sides the value of pi stays the same. This goes regardless of the distance of the revolving object from the center of the gravitational system, in this case the Sun. And farther way from the Sun, where the solar pressure exhibited is proportionally weaker, it will require less velocity to stay in orbit (smaller orbital velocity). This corresponds well with Kepler's third law of planetary motion.

Just as the Sun is said to have a gravitational field so are the planets, including the Earth. It is the means in which our laws of motion are said to be governed by in respect to the earth. All motion, change, and time is defined relative to and in respect to the earth. It is said to be the magnitude and nature of the earth's gravitational field by which random objects of a relatively significant mass remain at rest in respect to the earth, and fixed to its surface. The force of the field acting uniformly at all points on all objects within its physical range that affects and dictates motion in general and determines the physical trajectories of these same objects that are moved through the field. When dealing with the gravitational field of any given body, particularly stellar bodies such as stars, it is said that in the surrounding region outside of its surface as depicted in polar coordinates, the linear motion or geodesic of particles and stellar matter, deviates in proportion to the strength of the field. The linear motion then becomes curved, changing from its linear trajectory. The geodesic equation then varies within its new path and was described by the curvature tensor of space. In our new theory of gravity this is set at zero. We assume in general space to Euclidean throughout, homogenous and isotropic, and equivalent locally and globally. It is when we analyze space surrounding a massive stellar object, such as a star, that we find, locally the characteristics of space still to remain Euclidean. The linear motion of particles within a uniform space become curved taking on their new trajectories, in correlation to means by which they take on their new forms of motion. We can depict the trajectory of a classical particle in any given region of space and time using a vector, and the energy-momentum vector. It is within the framework of non-Euclidean geometry that we concern ourselves with the physical dynamics of material bodies and how their non-linear motion affects the non-local structure of space, all of which we disregard. That is the space surrounding a massive stellar body is still to remain Euclidean. If we neglect the construct of an external spacetime or spacetime continuum, as defined by the constant terms of the metric, and concern ourselves with the construct of Euclidean formation in 3-dimensional space, we find the curvature associated with the construct of the spacetime vanishes as does the values initially obtained from the Riemann Christoffel curvature tensor. If we place as our center of mass in a heliocentric system, defined as a Euclidean space consisting of three spatial dimensions, our Sun or stellar body, we will find along a designated elliptical plane within the defined space, a series of orbits as defined by the motion of a collection of bodies that are positioned within the heliocentric system. If along the defined heliocentric plane, we designate a divergent vector field as measured by the deviation of all nearby particles within range of the field, we will find the need for any curvature of physical space as superfluous in our description of stellar phenomenon within a heliocentric system. Then once the Riemann tensor and its associated connection coefficients are set at zero, we find use of the covariant derivative in our description of the relative deviation of all particles designated with the defined space, and their Newtonian curved trajectories essential. These trajectories are elliptical orbits orbiting about a center of gravity. We also have with the deviation of the particles within the heliocentric plane, in terms of their orbital trajectories, a family of curved paths distinguished by a set parameter n. We then also have a separation vector to measure the distance separating these trajectories. Taking our family of curved paths, within our heliocentric plane, we define their motion in respect to the construct of the heliocentric plane in terms of designated lines of force that define at each unit volume within the plane the strength of the field. By determining the separation vector and its parallel transport along the designated paths, we can determine the relative acceleration vector of the deviated paths defined by the orbital trajectory of the particles. The tidal gravitation forces we can define and measure by the deviation of the nearby family of particles. In correlation to the motion of the stellar bodies lying within the heliocentric plane and that of the particles within the field, we find the rotation of the Sun on its axis moving at a constant angular velocity, corresponds to the net output of quantified energy and radiation that emanates from the rotating body at a constant rate and perpendicular the direction of the orbiting bodies. This output combined with the input, that is the energy and material being pulled in by the sun in correlation to its intrinsic rotation and constant output, corresponds to the deviation of the vector field on the heliocentric plane. This also corresponds to the deviation and relative acceleration of the particles. As was defined, we then arrive at a motional description of the laws of gravity in determining the orbits of the planets in correlation to the state of the Sun as the center of the heliocentric system. But since the Riemann tensor is set at zero, the deviation of the test particles corresponds to the Newtonian equations for test bodies. The Newtonian constant as the cause of the gravitational force must be replaced by the deviation of the test bodies due to the construct of the vector field associated with the heliocentric system, coined as the gravitational radiation flux force, *G.F.F.F.*

2

Logical discrepencies associated with the theoretical and experimental foundations of general relativity

Introduction

Here when we mention that the two theories or principles are incompatible, we do not just mean that they disagree with each other, we imply that they both cannot coexist.

Special Relativity and General Relativity

If the Principle of Special Relativity holds, then General Relativity would predict circular orbits for orbiting bodies, and not elliptical orbits. Interestingly enough the shortest part of the orbit, the apogee, is always in line and in the front part of the orbit in relation to the direction of motion of the gravitational source, assuming in theory it is a spatially isolatory source (gravity is not generated by motion alone). Logically, the form of the gravitational field, according to General Relativity, along both four trigonometric radii, should be of equal length. It can not be due to time dilation, because it is not a directionally dependent phenomenon. Now if the speed of light varies in line with the direction of the moving gravitational body, then in front, in line with the moving source, the light should reduce in speed and does have lesser strength as it pushes against the the orbiting body. Now if the radiation, the moving light particles, were moving in an opposite direction to the moving gravitational body, then it would have a greater velocity in relation to the moving source, and thus have a greater force on the orbiting body, resulting in a decrease in distance between both bodies, in terms of the apogee, all proportional the magnitude of the moving gravitional body. Here we see that General Relativity and the Principle of Relativity are incompatible. This is then categorized as a frame-dependent phenomenon. We also pointed out that the principle of the constant velocity of light is in question as well.

General Relativity and Quantum Mechanics

One of the fundamental principles of *QM* is the wave-particle duality principle. This states that for each existing particle there has to be an associated wave. This results in two different ways. One, the singular particle exists both simultaneously as a wave and a particle. Second, the wave is transmitted via a medium of existing particles. Sound waves are incident waves. They are only transmitted by means of an existing medium of particles. They occur when two objects collide and require great strength in order to travel a great distance. They cannot be transmitted through empty space because there is no medium of particles, and if there were then they would have shown up as a fringe shift on the Michelson Interferometer experiment. When two black holes collided, some form of energy was released, possibly as gravitational waves. But in order for the waves to be transmitted through space, one of the prior conditions must be met. Thus these waves would have to consist of singular particles emitted by this collision, similar and different from sound waves. But if they are both singular and incidental waves why is it that gravitational waves and not sound waves can be transmitted through space when no medium of particles exist? Apparently all objects are to emit gravitational waves, but they are immeasurable. We can easily measure radiation emanating from larger objects, such as stars, and smaller objects such electrons. Black holes colliding is a singular event and not an ordinary object giving off radiation. They are, most likely, both incident waves and particles resulting in the process of objects colliding. Another problem involves the Strong Force or Nuclear Force, that binds protons to neutrons, and holds electrons in orbit. Knowing that electrons emit energy in the form of microscopic photons, how is it then, assuming that electrons and protons are of opposing charges, electrons can continually emit energy without running out and eventually exploding in the same manner as stars. The same goes for protons. How can they constantly absorb energy without imploding? Possibly, considering they are of significant different mass, they both hold the same charge, just of different magnitudes.

Miscellaenous

General Relativity cannot explain, as well, why some objects rotate in one direction clockwise, and others in the opposite direction, counterclockwise. How can a singular object curve space in two different directions? General Relativity cannot explan, likewise, why all orbiting objects rotate on their axes.

Experimental difficulties

Here we analyze the Pioneer Anomaly. If you have two forces, one moving perpendicular to the the, with radii extending directly from the gravitational source, and another parallel with the motion of the orbiting body, how can this body can stay in orbit. Considering the possibility that the condition of a curved medium, spacetime, can hold the object in orbit, counteracting a perpendicular force, then as the object moves away from the source, with both the gravitational field weakening, according to the inverse ratio principle, and the added force from the radiation emitted by the gravitational source, the object in question as it moves away, should be increasing in velocity instead of slowing down, thus the designation of an anomaly.

3

Tensor Analog

Introduction

Here we focus on the mathematical abstract construction of the field equations of the G.R.F.F. in terms of higher order tensors, and linear vector spaces.

Two vectors can be expressed in the component form

$$A = A_1 e_1 + A_2 e_2 + A_3 e_3 \quad \text{and} \quad B = B_1 e_1 + B_2 e_2 + B_3 e_3$$

here e_n are orthogonal unit basis vectors. We may also write

$$A = \left(A_1, A_2, A_3 \right) \tag{1.1}$$

only the components of the vector are given

The unit vectors can be represented as

$$e_1 = (1,0,0) : e_2 = (0,1,0) : e_3 (0,0,1) \tag{1.2}$$

In index notation, we will have

$$A_i, \rightarrow i = 1,2,3$$

In the vector space employed we will have three-dimensional, second order, accelerative vectors. This can be given as

$$dX / dt = \left((dX_1 / dt), (dX_2 / dt), (dX_3 / dt) \right) \tag{1.3}$$

Where N is the Newtonian constant. This can also be written as

$$dX / dt = \left((dX_1 / dt) e_1 + (dX_2 / dt) e_2 + (dX_3 / dt) e_3 \right) \tag{1.4}$$

where

$$e_1, e_2, ..., e_3 \tag{1.5}$$

represent linearly independent unit base vectors. The vector spaces employed represent systems or more specific, tensor systems.

When using tensors, the subscripts and superscripts are referred to as indices or suffixes. They are either lower case Latin or Greek letters. The number of these determine the order of the tensor system employed. Here we employ two indices which represent a second order system. There is also a difference in superscripts used and the powers of any quantity. To write a superscript quantity to a power we use parentheses.

The systems we employ here are symmetric because the superscripts when interchanged stay the same in terms of their values. It is also skew-symmetric since the tensor is unchanged when the components of the tensor change sign. This can be given as

$$\begin{aligned} v_1 &= a_{11} e_1 + a_{12} e_2 \\ v_2 &= a_{21} e_1 + a_{22} e_2 \end{aligned} \tag{1.6}$$

we use a dummy index in this case k. This can be given as

$$\begin{aligned} v_k &= a_{k1} e_1 + a_{k2} e_2 + a_{k3} e_3 \\ k &= 1, 2, 3 \end{aligned} \tag{1.7}$$

This can now be written as

$$v_k = \sum_{i=1}^{3} a_{ki} e_i = a_{k1} e_1 + a_{k2} e_2 \tag{1.8}$$

here is the dummy summation index. The subscript i repeats itself and the summation convention used requires that a summation be performed by allowing the summation subscript to take on the specific values by the range. Then we sum the results. The *k* index is called a free index.

We employ the summation convention when there is an mathematical expression in which an index occurs more than once on the same side of the equation. This represents a summation on the repeated indices. In terms of vector spaces, we employ a summation convention since the vectors are repeated in the same symmetric system. There is also a specified range when it comes to the summation convention used.

Calculus of Derivatives: Linear Vector Spaces

Fields:

These are special vector spaces of second order since they involve the partial differentiation of force vectors and thus fall under the category of special fields.

Each dual-special vector space is over an arbitrary field, denoted *F*, and another field denoted *2F*. This is a set together with four binary operations, two forms of addition and multiplication. The first is of course vector addition and scalar multiplication that satisfy the regular axioms, already assumed. This includes:

associativity of addition

commutativity of addition

identity element of addition

inverse element of addition

distributivity of scalar multiplication in terms of vector addition

distributivity of scalar multiplication in terms of field addition

Compatibility of scalar multiplication with field multiplication

Identity element of scalar multiplication

This is a special set of dual-base vector spaces, as defined by the divergence and convergence of each vector as defined in terms of coordinates in a dual-converging and diverging vector fields. Each vector space is over a special field *K* as a set *V* together with two new binary operations. The other addition operation involves the pairing of each vector in each vector space with an arbitrary value involving the element of pi, which involves one special field. The other multiplicative operation involves the pairing of each vector with the element denoted *M*, that signifies the mass in question, the other special field in question. This is the *G.R.F.F.* acting on the orbiting body in question. Each vector in these vector spaces are forces themselves that constitute the *G.R.F.F.* It satisfies the following axioms

Associativity of addition

Commutativity of addition

Identity element of addition

Inverse element of addition

Distributivity of scalar multiplication in terms of pi and vector addition

Compatibility of scalar multiplication with field multiplication, in terms of M.

Identity element of scalar multiplication of M.

The converging and diverging tensors, that consist of the singular gravitational field for every gravitational source, are linear vector spaces themselves.

Linear Transformations:

Given the dual vector spaces over a field, given as the generalized field *G*, a linear transformation is a map given as

$$M : V \to W \qquad\qquad (1.9)$$

where we have the compatibility of addition and scalar multiplication

$$M(u+v) = Mv + Mu$$
$$M(uv) = uM(v)$$

(1.10)

where for any two vectors $u, v \in V$ and scalar $a \in F$.

Subspaces, Span, and Basis:

The set of all linear combinations of vectors $v_1, v_2, ..., v_k$ is referred to as their span, which is in itself a subspace. Since some vectors are linear combinations of other, the set is linearly dependent. The basis of a vector space is the linear dependent set of vectors that span the vector space. The cardinality of bases of each vector space is also referred to as the dimension of *V*, which in this case is two. Since the basis of each special vector space has a finite number of elements, it is referred to as a finite-dimensional vector space. Since the span of vectors in each vector space is a subspace of the main vector spaces *V* and *U*, then in terms of the subspaces *u* and *v* we will have

$$\dim(u_1 + u_2) = \dim u_1 + \dim u_2 - \dim(u_1 \cap u_2)$$

(1.11)

Here we employ the chain rule for the differentiation of a function of the bar variables. Here we use the indicial notation. Let the normal vector be given as $V = V(dx_1, dx_2, ..., dx_N)$. Here *N* is the Newtonian constant. The partial derivatives of *V* with respect to the variables of *x* can be given in terms of the indicial notation as

$$\frac{\partial^2 V}{\partial x_i} = \frac{\partial V}{\partial x_j}\frac{\partial x_j}{\partial x_i} = \frac{\partial V}{\partial x_j}\frac{\partial x_j}{\partial x_i} + \frac{\partial V}{\partial x_j}\frac{\partial x_j}{\partial x_i} + ... + \frac{\partial V}{\partial x_N}\frac{\partial x_N}{\partial x_i}$$

(1.12)

This can represent a singular vector space as

$$\left\{ \sum_{i=1}^{3}\left(\frac{\partial^2 V}{\partial x_i}(\pm)\sin + \frac{\partial^2 V}{\partial x_i}(\pm)\cos \right) + \frac{\partial^2 V}{\partial x_i}\sin + \sum_{i=1}^{3}\left(\frac{\partial^2 V}{\partial x_i}(\pm)\sin + \frac{\partial^2 V}{\partial x_i}(\pm)\cos \right) \right\}\frac{Mm}{r^2}$$

(1.13)

Now let $V = V(r,\theta)$ where r and θ are polar coordinates related to the Cartesian coordinates (x,y) by the transformation equations $x = r\cos\theta : y = \sin\theta$. The partial derivates of the ordinary vectors may be given as

$$\frac{\partial^2 V}{\partial x_i^2} = \left\{ \left(\frac{\partial V}{\partial r_1}\frac{\partial^2 r}{\partial x^2} + \frac{\partial r_1}{\partial x}\frac{\partial}{\partial x}\left[\frac{\partial V}{\partial r_1}\right] \right) + \left(\frac{\partial V}{\partial \theta_1}\frac{\partial^2 \theta_1}{\partial x^2} + \frac{\partial \theta_1}{\partial x}\frac{\partial}{\partial x}\left[\frac{\partial V}{\partial \theta_1}\right] \right) \right\}$$
$$+ \left(\left(\frac{\partial V}{\partial r_2}\frac{\partial^2 r_2}{\partial x^2} + \frac{\partial r_2}{\partial x}\frac{\partial}{\partial x}\left[\frac{\partial V}{\partial r_2}\right] \right) + \left(\frac{\partial V}{\partial \theta_2}\frac{\partial^2 \theta_2}{\partial x^2} + \frac{\partial \theta_2}{\partial x}\frac{\partial}{\partial x}\left[\frac{\partial V}{\partial \theta_2}\right] \right) \right)$$
$$+ ... + \left(\left(\frac{\partial V}{\partial r_N}\frac{\partial^2 r_N}{\partial x^2} + \frac{\partial r_N}{\partial x}\frac{\partial}{\partial x}\left[\frac{\partial V}{\partial r_N}\right] \right) + \left(\frac{\partial V}{\partial r_N}\frac{\partial^2 r_N}{\partial x^2} + \frac{\partial r_N}{\partial x}\frac{\partial}{\partial x}\left[\frac{\partial V}{\partial r_N}\right] \right) \right)_N \qquad (1.14)$$

and

$$\left\{ \sum_{i=1}^{3}\left(\frac{\partial^2 V}{\partial x_i^2}(\pm)\sin + \frac{\partial^2 V}{\partial x_i^2}(\pm)\cos \right) + \frac{\partial^2 V}{\partial x_i^2}\sin + \sum_{j=1}^{3}\left(\frac{\partial^2 V}{\partial x_j^2}(\pm)\sin + \frac{\partial^2 V}{\partial x_j^2}(\pm)\cos \right) \right\}\frac{Mm}{r^2} \qquad (1.15)$$

To simplify this equation even further, the terms inside the bracket are given as functions of the variables r and θ and the derivative of these terms are evaluated by reapplying the rule from equation (1.11). This can be given as

$$\frac{\partial^2 V_i}{\partial x_i^2} = \left\{ \begin{array}{l} \dfrac{\partial V}{\partial r_1}\dfrac{\partial^2 r_1}{\partial x^2} + \dfrac{\partial r}{\partial x}\left[\dfrac{\partial^2 V}{\partial r_2^2}\dfrac{\partial r_1}{\partial x} + \dfrac{\partial^2 V}{\partial r\partial\theta}\dfrac{\partial\theta}{\partial x}\right] \\[3mm] + \dfrac{\partial V}{\partial \theta_1}\dfrac{\partial^2 r}{\partial x^2} + \dfrac{\partial\theta}{\partial x}\left[\dfrac{\partial^2 V}{\partial r_1\partial\theta}\dfrac{\partial r}{\partial x} + \dfrac{\partial^2 V}{\partial\theta^2}\dfrac{\partial\theta}{\partial x}\right] \end{array} \right\}_1$$
$$+ \left\{ \begin{array}{l} \dfrac{\partial V}{\partial r_2}\dfrac{\partial^2 r_2}{\partial x^2} + \dfrac{\partial r}{\partial x}\left[\dfrac{\partial^2 V}{\partial r_3^2}\dfrac{\partial r}{\partial x} + \dfrac{\partial^2 V}{\partial r_1\partial\theta}\dfrac{\partial\theta}{\partial x}\right] \\[3mm] + \dfrac{\partial V}{\partial \theta_2}\dfrac{\partial^2 r_2}{\partial x^2} + \dfrac{\partial\theta}{\partial x}\left[\dfrac{\partial^2 V}{\partial r_2\partial\theta}\dfrac{\partial r_2}{\partial x} + \dfrac{\partial^2 V}{\partial\theta^2}\dfrac{\partial\theta}{\partial x}\right] \end{array} \right\}_2 \qquad (1.16)$$
$$+ ... + \left\{ \begin{array}{l} \dfrac{\partial V}{\partial r_N}\dfrac{\partial^2 r_N}{\partial x^2} + \dfrac{\partial r}{\partial x}\left[\dfrac{\partial^2 V}{\partial r_{N+1}^2}\dfrac{\partial r}{\partial x} + \dfrac{\partial^2 V}{\partial r\partial\theta}\dfrac{\partial\theta}{\partial x}\right] \\[3mm] + \dfrac{\partial V}{\partial \theta_N}\dfrac{\partial^2 r}{\partial x^2} + \dfrac{\partial\theta}{\partial x}\left[\dfrac{\partial^2 V}{\partial r_N\partial\theta}\dfrac{\partial r}{\partial x} + \dfrac{\partial^2 V}{\partial\theta^2}\dfrac{\partial\theta}{\partial x}\right] \end{array} \right\}_3$$

Then given the following relations

$$2r\frac{\partial r}{\partial x} = 2x \rightarrow \frac{\partial r}{\partial x} = \frac{x}{r} = \cos\theta$$

$$\sec^2\theta\frac{\partial \theta}{\partial x} = -\frac{y}{x^2} \rightarrow \frac{\partial \theta}{\partial x} = -\frac{y}{r^2} = -\frac{\sin\theta}{r}$$

$$\frac{\partial^2 r}{\partial x^2} = -\sin\frac{\partial \theta}{\partial x} = \frac{\sin^2\theta}{r}$$

$$\frac{\partial^2 \theta}{\partial x^2} = \frac{-r\cos\dfrac{\partial \theta}{\partial x} + \sin\theta\dfrac{\partial r}{\partial x}}{r^2} = \frac{2\sin\theta\cos\theta}{r^2}$$

(1.17)

The derivates given before can then be written as

$$\frac{\partial^2 \Theta}{\partial x^2} = \frac{\partial \Theta}{\partial r}\frac{\sin^2\theta}{r} + 2\frac{\partial \Theta}{\partial \theta}\frac{\sin\theta\cos\theta}{r^2} + \frac{\partial^2\Theta}{\partial x^2}\cos^2\theta - 2\frac{\partial^2\Theta}{\partial r\partial\theta}\frac{\sin\theta\cos\theta}{r^2} + \frac{\partial^2\Theta}{\partial\theta^2}\frac{\sin^2\theta}{r^2}$$

(1.18)

from this we can derive the following equation

$$\sum_{i=1}^{3}\left(\frac{\partial^2 \Theta}{\partial x^2}\right)_i = \begin{pmatrix} \dfrac{\partial \Theta}{\partial r}\dfrac{\sin^2\theta}{r} + 2\dfrac{\partial \Theta_1}{\partial \theta}\dfrac{\sin\theta\cos\theta}{r^2} + \dfrac{\partial^2\Theta_1}{\partial x^2}\cos^2\theta \\ -2\dfrac{\partial^2\Theta_1}{\partial r\partial\theta}\dfrac{\sin\theta\cos\theta}{r^2} + \dfrac{\partial^2\Theta_1}{\partial\theta^2}\dfrac{\sin^2\theta}{r^2} \end{pmatrix}$$

$$+ \begin{pmatrix} \dfrac{\partial \Theta_2}{\partial r}\dfrac{\sin^2\theta}{r} + 2\dfrac{\partial \Theta_2}{\partial \theta}\dfrac{\sin\theta\cos\theta}{r^2} + \dfrac{\partial^2\Theta_2}{\partial x^2}\cos^2\theta \\ -2\dfrac{\partial^2\Theta_2}{\partial r\partial\theta}\dfrac{\sin\theta\cos\theta}{r^2} + \dfrac{\partial^2\Theta_2}{\partial\theta^2}\dfrac{\sin^2\theta}{r^2} \end{pmatrix}$$

$$+\ldots+ \begin{pmatrix} \dfrac{\partial \Theta_N}{\partial r}\dfrac{\sin^2\theta}{r} + 2\dfrac{\partial \Theta_N}{\partial \theta}\dfrac{\sin\theta\cos\theta}{r^2} + \dfrac{\partial^2\Theta_N}{\partial x^2}\cos^2\theta \\ -2\dfrac{\partial^2\Theta_N}{\partial r\partial\theta}\dfrac{\sin\theta\cos\theta}{r^2} + \dfrac{\partial^2\Theta_N}{\partial\theta^2}\dfrac{\sin^2\theta}{r^2} \end{pmatrix}$$

(1.19)

The field equations can then be given as

$$\left\{ \sum_{i=1}^{3}\left(\frac{\partial^2 \Theta}{\partial x_i^2}(\pm)\sin\theta + \frac{\partial^2 \Theta}{\partial x_i^2}(\pm)\cos\theta\right)_N + \frac{\partial^2\Theta}{\partial x_i^2}\sin\theta \right|_N \\ + \sum_{j=1}^{3}\left(\frac{\partial^2 \Theta}{\partial x_j^2}(\pm)\sin\theta + \frac{\partial^2 \Theta}{\partial x_j^2}(\pm)\cos\theta\right)_N \right\} \frac{Mm}{r^2}$$

(1.20)

Each vector space can also be defined mathematically as a divergence of a vector field in Cartesian coordinates represented as

$$(\nabla \cdot A) = (divA) = A_{ij} = \frac{\partial A_i}{\partial x_i} = \frac{\partial A}{\partial x_1} + \frac{\partial A}{\partial x_2} + \frac{\partial A}{\partial x_3}$$

$$\begin{Bmatrix} \left[\sum_{i=1}^{3} \left(\frac{\partial A}{\partial x}(\pm)\sin\theta + \frac{\partial A}{\partial x_1}(\pm)\cos\theta \right) + \frac{\partial A}{\partial y}\sin\theta \right]_N \\ + \sum_{j=1}^{3} \left(\frac{\partial A}{\partial z}(\pm)\sin\theta + \frac{\partial A}{\partial z}(\pm)\cos\theta \right)_N \end{Bmatrix} \frac{Mm}{r^2} \tag{1.21}$$

Integral Theorems

In terms of the divergence theorem in both vector and indicial notation we will have

$$\iint_S (\nabla \times F) \cdot n \, d\sigma + \int_C F \cdot dr \rightarrow \int_S e_{ijk} F_{jk} n_i \, d\sigma = \int_C F_i \, dx^i \tag{1.22}$$

$$i, j, k = 1, 2, 3$$

where *F* are components of a generalized force.

Tensors:

Given the independent orthogonal unit vectors (base vectors) we may write a random vector as

$$A = A_1 e_1 + A_2 e_2 = (A \cdot e_1) e_1 + (A \cdot e_2) e_2 = \left(\frac{A \cdot E_1}{E_1 \cdot E_1} \right) E_1 + \left(\frac{A \cdot E_2}{E_2 \cdot E_2} \right) E_2 = \left(\frac{A \cdot E_i}{E_i \cdot E_i} \right) E_i \tag{1.23}$$

$$i = 1, 2$$

Here the three-dimensional converging and diverging vector fields coupled to every gravitational source, are third-degree tensors.

Reciprocal Basis

We will start with a set of two independent vectors (E_1, E_2) of a unit length. We can represent the vector A in terms of these above listed vectors as

$$A = A^1 E_1 + A^2 E_2 \tag{1.24}$$

Given the vector r will we then have

$$\frac{d^2r}{dt^2} = \frac{d^2x}{dt^2}\hat{e}_1 + \frac{d^2y}{dt^2}\hat{e}_1 = \frac{d^2r}{du^2}d(du) + \frac{d^2r}{dv^2}d(dv) \tag{1.25}$$

where

$$\frac{d^2r}{du^2} = \frac{d^2x}{du^2}\hat{e}_1 + \frac{d^2y}{du^2}\hat{e}_1$$
$$\frac{d^2r}{dv^2} = \frac{d^2x}{dv^2}\hat{e}_1 + \frac{d^2y}{dv^2}\hat{e}_1 \tag{1.26}$$

In terms of the u and v coordinates, this acceleration vector has vector sides.

Assume than $u = u(x,y)$, we will have

$$\frac{d^2u}{dx^2} = \frac{d^2u}{dx^2}dx + \frac{d^2u}{dy^2}dy \tag{1.27}$$

This differential can be written as

$$\frac{d^2u}{dx^2} = gradu \cdot \frac{d}{dt}(dr) = gradu\left(\frac{d^2r}{du^2}du + \frac{d^2r}{dv^2}dv\right)$$
$$= \left(gradu \cdot \frac{d^2r}{du^2}du\right) + \left(gradv \cdot \frac{d^2r}{dv^2}dv\right) \tag{1.28}$$

Dyads and Polyads: Tensors

If we can represent vectors as

$$A = A_i E^i \tag{1.29}$$

This notation can be used when representing tensor quantities. In terms of basis vectors a second order tensor can be given as

$$\Gamma = T_{ij} E^i E^j = T_{ij} M^i T^j \tag{1.30}$$

Here M and T are called dyads. A dyad tensor is formed by the tensor product of two tensors. In Cartesian coordinates, we will have $T^i = T_i = \hat{t}_i$ and $M^i = M_i = \hat{m}_i$. We will then have

$$T = T_{11}\hat{t}_1\hat{m}_1 + T_{12}\hat{t}_2\hat{m}_1 + T_{21}\hat{t}_1\hat{m}_1 + T_{22}\hat{t}_2\hat{m}_1 \tag{1.31}$$

Adding Tensors:

Tensors of the same weight and type can be added. In terms of two similar tensors when added we will have

$$A^{ij} = C^{ij} + B^{ij} \tag{1.32}$$

which in matrix form can be read as

$$A^{ij} = B^{ij} + \begin{bmatrix} \pm \dfrac{\alpha}{\beta}\pi \\[2ex] \pm \dfrac{\alpha}{\beta}\pi \end{bmatrix} \tag{1.33}$$

Tensor Notation

Gradient:

For $\Theta = \Theta\left(v^1, v^2, ..., v^N\right)$ the vectors of which the gradient is defined as covariant force vectors, we will have

$$\Theta_i = \frac{\partial \Theta}{\partial x^i} : i = 1, ..., N \tag{1.34}$$

Divergence

The divergence of a contravariant tensor *A* is given by

$$divA^r = A^r_{,r} \qquad (1.35)$$

The divergence of a vector *A* in spherical coordinates is given as

$$\nabla \left(dA^r / dx \right) = \frac{1}{\sqrt{g}} \left[\frac{\partial}{\partial x^1} \left(\sqrt{g} A^1 \right) + \frac{\partial}{\partial x^2} \left(\sqrt{g} A^2 \right) \right] \qquad (1.36)$$

Generalized force as generalized acceleration

By chain rule differentiation, the generalized acceleration vector may be given as

$$F^r = \frac{dv^r}{dt} = v \frac{dT^r}{dt} = v \frac{dT^r}{dt} \frac{ds}{dt} \qquad (1.37)$$

Tensor Product

The tensor product of two three-dimensional vector spaces coupled to another tensor, the mass tensor (energy-mass tensor) all constitute the generalized G.R.F.F. tensor. *Here each gravitational body is endowed with two third-degree tensors, three -dimensional diverging and converging vector spaces.* It is denoted as

$$(V \otimes V)M \qquad (1.38)$$

The normal definition of the tensor product symbol involves the idea of two vector spaces, denoted, $F(S)$ and . The formal sums of the elements *S, vectors,* has its coefficients in each ordinary tensor. Given two vector spaces and another tensor denoted *M*, the product, given in (1.38), is the G.R.F.F. field. The necessary elements and operations are given. From the cartesian product

$(V \otimes U)M$, the free vector space denoted as $F(V \otimes U)M$. The vectors of one vector space and the related elements of the other arbitrary space, will be defined by the equivalence classes of $F(V \otimes U)M$ corresponding to the following equivalence relations

$$
\begin{aligned}
&v \in V : u \in U : m \in M : c \in: \\
&(v_1, m) + (v_2, m) = (v_1 + v_2, m) : (v, m_1) + (v, m_2) = (v, m_1 + vm_2): \\
&(u_1, m) + (u_2, m) = (u_1 + u_2, m) : (u, m_1) + (u, m_2) = (u, m_1 + um_2): \\
&c(v, m) = (cv, m) = (v, cm) \\
&c(u, m) = (cu, m) = (u, cm)
\end{aligned}
\tag{1.39}
$$

The tensor product given above is defined as the quotient space $(F(V \otimes U)M)/N$ over the subspace N of $F(V \otimes U)M$

We will then have

$$
N = \begin{bmatrix}
n \in F(V \otimes U)M : \exists (v_1, v_2 \in V \wedge u_1, u_2 \in U \wedge m_1, \in M \wedge c \in K) \\
n = (v_1, m_1) + (v_2, m_1) - (v_1 + v_2, m_1)v \\
n = (u_1, m_1) + (u_2, m_1) - (u_1 + u_2, m_1)v \\
n = c(v_1, m_1) - (cv_1, m_1) \vee n = c(v_1, m_1) - (v_1, cm_1) \\
n = c(u_1, m_1) - (cu_1, m_1) \vee n = c(u_1, m_1) - (u_1, cm_1)
\end{bmatrix}
\tag{1.40}
$$

Given the basis $\{v_i\} : \{u_j\} : \{m\}$ for V and U and M, the tensors $\{v_i \otimes u_j\}m$ form a basis for $V \otimes U(M)$. The dimension of the tensor product is the product of the dimensions of the original spaces. The elements of the product of the tensor afore mentioned are referred to as tensors as well. An element of this product is referred to as a pure or simple tensor. It is a finite linear combination of pure tensors. Elements from both tensors are linearly independent. The number of simple tensors required to express an element of a tensor product is called the tensor rank, which in this case is directly proportional to the Newton constant, N.

Just as the linear vector space of dimension *N*, proportional to Newton's constant, couples to the arbitrary tensor, denoted *M*, there are is also an isomorphic vector space that couples to the same arbitrary tensor, *M*. All of this is now given as the *G.R.F.F.* Tensors.

We can draw a linear map on which the tensor product operates. This can be given as $S:V \to M$ and $T:W \to M$, between two vector spaces that couple, are then also the direct tensor product of the tensor *M*. This can be given as

$$(S \otimes T):V \otimes W \to M \tag{1.41}$$

defined by

$$(S \otimes T)(M) = S(M) \otimes T(M) \tag{1.42}$$

this tensor product is defined as a bifunctor in terms of vector spaces, covariant in both arguments. As long as both *S* and *T* are injective, surjective and continuous, then so is their tensor product. In terms of the vector space bases, the linear maps *S* and *T* can also be represented by matrices. The tensor product of both *S* and *T* is the known as the Kronecker product of both matrices.

The tensor product is also symmetric, meaning there is a canonical isomorphism. The second tensor power of the main tensor *P* is a result of the summation of double product of two isomorphic vector spaces and the arbitrary tensor *M*. This, in terms of a (r,s) tensor on a two dual vector spaces *V* and *W,* is an element of

$$T_s^r(V) = (V \otimes M) \oplus (W \oplus M) = V^{\otimes r} \oplus W^{\otimes s} \tag{1.43}$$

where

$$W = V \tag{1.44}$$

here *W* can also be considered to be a dual vector space to *V*, adjoined by the main tensor *M*. Choosing a basis of *V* and the corresponding dual basis of *V*, *W*, $T_s^r(V)$ is endowed with a normal basis. The normalized general tensor, as a computation of combining two ordinary second order tensors, in this case dual vector spaces with a singular generalized tensor denoted *M*, can be defined by the following master equation

$$\Gamma_i^{uv\oplus av} = \left(V^u M^v\right)\oplus\left(W^u M^v\right)=\left(V\otimes M\right)^{uv}\oplus\left(W\otimes M\right)^{av} \tag{1.45}$$

Fields

The main operator group *G* is also a special field and ring that is a combination of two isomorphic subgroups that are fields themselves, consisting of at set, given as *K*. In which there are both the addition and multiplication operations. They satisfy the following axioms:

- $V_N + \left(\pm\pi\,\frac{\alpha}{\beta}\right)=\left(\pm\pi\,\frac{\alpha}{\beta}\right)+V_N$ for all elements of *R* (commutative); where α and β are arbitrary constants and *V* is the main generalized element for each linear vector space of a given subgroup. N is the Newtonian constant. This will be replaced by the *G.R.F.F.* constant. Here the arbitrary value of pi couples to every vector in the linear vector space, making up a group or subgroup to the main operator group, denoted *G*.

- This ring is not associative in terms of addition of its elements.

- There exists an element 0 known as the zero element with the property *x+0*.

Given an element of *K* there is the inverse element to that element with the property $V_N +(-V_N)=0$

$V_N M = MV_N$ for all elements of K (multiplication is communitive).

Multiplication is not associative.

The distributive law does not apply.

4

The field equations: group theory

Introduction

The gravitational radiation flux field (G.R.F.F.) is also an Abelian Operator Group, denoted G. It is a set with an multiplative operation denoted by the symbol $(G \cdot)$.

This multiplative operation couples two dual isomorphic vector spaces to the same value or scalar quantity denoted M. The nature of the *G.R.F.F.* dictates the behavior of *m,* operating as a main operator group in terms of group theory as it couples *m* to the two vector spaces. That is to say, *G* is the group of all force vectors, given as elements of the main set representing *G,* that couples to the variable denoted *M*, whose product represents the quantity *w*, referred to as the angular momentum. Both subgroups must have the same number of vectors, second-order vectors, and be positioned at the same angle for each mirror vector, between both subgroups of vectors, thusly producing *w*.

Generally given an arbitrary scalar value denoted, *m*, is defined so because it has no direction as vectors generally do. The group *G* takes a finite number of vectors, elements of the group, and couples then in such a specific manner to *m*, so as to produce a new vector, the angular momentum vector, denoted *w*.

This is known as the group law. It combines any two elements to form a third element. Group *M* is the combination of two connected and symmetric linear second order vector spaces. These are also subgroups of the main operator group *G* and groups themselves.

G is a group because it satisfies the four axioms of group theory.

a. The first is the closure law, in which the combining of the elements from both vector spaces with the element *M*, the mass of the body in question, are both found in the operator group, denoted *G*.

b. The second axiom is the law of associativity, in which all elements of the subgroups, part of *G*, couple in the same manner to *M* regardless of the order.

c. There also exists an identity element.

d. The final axiom is the existence of an inverse element, for every element in *G*.

Subgroups:

We begin with the main operator group, denoted again as *G*. *H* is the subset of *G*. *H* is also a subgroup of *G* according to the same conditions applied to the generalized operator group *G*. This subgroup is proper as long as $H \neq G$. Because the elements of *G*, in the subgroups are of such form $\dfrac{d^2 x_\alpha}{dx_\beta{}^2}$, the group *G* with generator *(dx/dt)*, is defined as a cyclic group. Let *H* be a subgroup of *G*, then the left and right coset of *H* in *G* is a subset of *G* in the form $\dfrac{d^2 x_\alpha}{dx_\beta{}^2}$ H, where $\dfrac{d^2 x_\alpha}{dx_\beta{}^2} \in G$ and

$$\frac{d^2 x_\alpha}{dx_\beta{}^2} H = \left\{ \frac{d^2 y_\alpha}{dy_\beta{}^2} \in G : \frac{d^2 y_\alpha}{dy_\beta{}^2} = \frac{d^2 x_\alpha}{dx_\beta{}^2} h \rightarrow for; h \in H \right\}$$

and the right coset is given as

$$H \frac{d^2 x_\alpha}{dx_\beta{}^2} = \left\{ \frac{d^2 y_\alpha}{dy_\beta{}^2} \in G : \frac{d^2 y_\alpha}{dy_\beta{}^2} = h \frac{d^2 x_\alpha}{dx_\beta{}^2} \rightarrow for; h \in H \right\}$$

Theorem: If *G* is finite group and *H* is a subgroup of *G*, then the order of *H* divides the order of *G*.

Theorem: If H is a subgroup of G and the number of left and right cosets of H in G is finite, then the number of such cosets is termed as the index of H in G, given by $[G:H]$.

Cor. Let $\dfrac{d^2 x_\alpha}{dx_\beta^{\ 2}}$ be an element of the finite operator group G, then the order of $\dfrac{d^2 x_\alpha}{dx_\beta^{\ 2}}$ divides the order of G.

Normal subgroups:

Let A, B, and M be subsets of group G. The product of AM and BM of the sets A, B, and M is given as

$$AM = \left\{ \left(\frac{d^2 x_\alpha}{dx_\beta^{\ 2}} \frac{d^2 y_\alpha}{dy_\beta^{\ 2}} \right) m : \left(\frac{d^2 x_\alpha}{dx_\beta^{\ 2}} \frac{d^2 y_\alpha}{dy_\beta^{\ 2}} \right) \in A \to m \in M \right\}$$

$$BM = \left\{ \left(\frac{d^2 x_\sigma}{dx_\tau^{\ 2}} \frac{d^2 y_\sigma}{dy_\tau^{\ 2}} \right) m : \left(\frac{d^2 x_\sigma}{dx_\tau^{\ 2}}, \frac{d^2 y_\sigma}{dy_\tau^{\ 2}} \right) \in B \to m \in M \right\}$$

AM and BM are interchangeable. Then if A, B, and M are subsets of the group G then we will have $A \subset G$, $M \subset G$ and $B \subset G$. We will also have $B \subset G$.

General definition of main operator froup G:

Each vector space of second order is also a subgroup and group in itself. It has four main operations in respect to the main generalized operator group, denoted G also referred to as the *G.R.F.F.* group. There are two additions and two multiplicative operations. The first two are referred to as classical operations and the latter two as special operations. Each vector space their vectors are added to form new vectors and the also a scalar multiplication. The special operations involve the arbitrary value of pi that is added to each vector in the vector space and the latter multiplicative operation the value of *M*, the body under the influence of the gravitational field, where each vector is coupled to.

Group *M* is also a generalized permutation group, since the acceleration vectors in both linear vector spaces are elements of two symmetric subgroups, denoted *g1 and g2*. Both subgroups are permutations groups themselves. The order of elements is symmetric with respect to both subgroup permutation groups. Each subgroup has elements themselves, vectors in terms of vector spaces, that form a dual-symmetric permutations of a set, denoted *X*. So, in terms of the subgroups, we will have two symmetric sets, denoted *X1* and *X2*. And they are equal. They are always finite sets. The number of elements or the modality of the set and its order of permutations is in direct ratio to Newton's constant, replaced with the *G.R.F.F.* We can define the gravitational field in terms of a generalized gravitational field within a two-dimensional phase space embedded geometrically within a three-dimensional Euclidean space. This we term mathematically as the *gravitational energy flux force (G.R.F.F.)*. The group operation of *M*, that defines *M*, is the composition of two sets of elements, the acceleration vectors themselves, their permutations defined in ration to *N* and the constant and arbitrary value of pi. The values of pi form a group themselves. So, the permutations in each subgroup is coupled to the subgroup, formed by the value pi. The coupling can be defined in terms of group theory as formation of the permutations of each subgroup as the composition of its elements with the elements of the main group *P* with the same dimensionality and order as the subgroups. This implies that each vector space is the set of elements, acceleration vectors, that couple with the set of elements of the group *P*. These are also thought of as bijective functions from the main set of each subgroup, to set off the main operator group *P*. The order of permutations in each set is in ratio to the values obtained from the *G.R.F.F.* The order is always symmetric but the total numbers varies depending on *G.R.F.F.* The permutations order and structure is always directly symmetric for both sets in G, the main operator group. Because of this, the main operator group G is also a symmetric group. Each separate permutation group, as the composition of elements from the set of group *P* and subgroups *g1* and *g2*, are subgroups themselves. Since *g1* and *g2* are permutations groups on the same set X, they then are classified as isomorphic as permutation groups. They are a bijective map between both, such that for each element in $g1$ there is also an equivalent element in $g2$. This implies than that both $g1$ and $g2$ are subgroups of G. Than from this we can derive the equality $Sym(X)=G$.

Group $G2$ is coupled in terms of multiplication to the main operator group M. This is also associative. The set of elements of $G2$ is coupled to the same number of elements defined in group M. For each group, there is also an element deemed as the identity element. Each group is also a finite group. The order of the group is in direct ration to the value of G.R.F.F. We can define the G.R.F.F. matrix mathematically in terms of two symmetric parts. It has a total number of elements in the matrix in proportion to the original Newton constant. We couple with the values of each element in the matrix with the arbitrary but constant value of pi. In terms of group theory, it is a doubly divergent n-dimensional vector space with symmetric duality. Due to the duality, the group G acting on its set of elements X when split into two subgroups $g1$ and $g2$, are all bijections or permutations from G onto both subgroups. They must be symmetric or equal. These are specifications of something more general, known as a transformation group. It is also a symmetric group.

A group can be defined in terms of our linear vector space as

$$\left(S_i, S_j\right) \boxtimes S_i * S_j : G \times G \rightarrow G$$

Both subgroups, $g1$ and $g2$, are isomorphic to each other. Just as long as there exists a one-to-one correspondence $S_i \leftrightarrow S'_i, G \leftrightarrow G'$ such that $\left(S_i * S_j\right) = S'_i * S'_j$ for all $S_i, S_j \in G$. We now focus on permutation groups. Here we have general group G and the Sym(G) group, with their set of bijections $S_i, S_j : G \rightarrow G$. Than the product of the two elements of sym(G) to be composite $S_i m \oplus S_j m = \left(S_i \boxtimes\right) \oplus \left(S_j \ m\boxtimes\right)$. Than Sym(G) is the group of symmetries of G. G is the result of the direct product of subgroups $g1$ and $g2$ coupling to group M.

Group *G* is also commutative or abelian. The order of group *G* or its cardinality is of the second order. As long as the order of group *G* is of the power prime *p*, it will be labeled as a *p* group as well. The general linear group *G*, of degree *n,* is defined. Than we have the two dual *F*-vector spaces, denoted as *V's*. Let *V* be a finite dimensional vector space over a field *F*, where in terms of group theory will be the main operator group *G*. A bilinear form on *V* is a mapping $a:V \times V \to F$ that is actually linear with respect to each variable. The automorphism is also an isomorphism. Both *g1* and *g2* are nonempty subsets of group *G* and then the binary operation on G makes both *g1* and *g2* into groups and subgroups of *G*. The linear mapping from the main group *M* onto the operator main group *G,* will consist of the symmetric set of mappings from *M*, from each element of that group denoted *m*, onto the subgroups of *G*, denoted *g1* and *g2.* This is also deemed as dual linear vector spaces with dimensionality *n.*

It is a doubly split vector space in which we generalize as a divergent group with *n*-dimensionality. The elements of the group are parts of a subgroup consistent of symmetrically divergent force-vectors. In terms of group theory, it is finite linear algebraic group. In terms of operations in the group and mathematically in the matrix we have two basic operations, addition and multiplication. In terms of addition we have the group operation that couples with each element the factor of pi. In terms of multiplication we have the coupling of each element with *m*, which non-mathematically would be the mass or body in question. It is also trigonometrically defined along the divergent two-dimensional phase space in terms of sin and cosine functions in a two-dimensional geometric coordinate system.

We define the main operator group *G* as

$$G \rightarrow g_1 + g_2 = \left(S_1(1,2), S_2, S_3(1,2) \right) + \left(S_4(1,2), S_5, S_6(1,2) \right)$$

$$= \begin{pmatrix} \left(S_1(1,2), S_2, S_3(1,2) \right) \\ \left(S_4(1,2), S_5, S_6(1,2) \right) \end{pmatrix} = \left(f(x,y, f(y), f(x,y)) \right) \qquad (1.46)$$

$$+ \left(f(x,y, f(y), f(x,y)) \right) = \begin{pmatrix} \left(f(x,y, f(y), f(x,y)) \right) \\ \left(f(x,y, f(y), f(x,y)) \right) \end{pmatrix}$$

Since group G is a linear vector space or more generally an algebra, we will have

$$G \rightarrow \sum_{i=1}^{h} g_i m_i \sum_{j=1}^{h} g_j m_i = \sum_{ij=1}^{h} g_{ij} m_i$$

and we have the additive operation of main linear operative group G, the subgroups g1 and g2 symmetric union, onto the arbitrary group P, denoted by the value of pi.

5

Generalized form of field equations in terms of ordinary differential equations

Introduction

Here we will formulate the actual field equations of the G.R.F.F. This is done just in terms of the ordinary differential equations, by combining vectors as previously derived in terms of special linear vector spaces.

In terms of the ordinary field equations written in terms of ordinary linear differential equations, we will have

$$F(x,y) = P(x,y)i + Q(x,y)j \qquad (1.47)$$

but since these are force vectors, rather than ordinary vectors, we will have in terms of an ordinary differential equations

$$\left\{ \begin{array}{l} \left(\sum_{i=\frac{Mm}{r^2}}^{m} A(-x,-y) + B(-y) + \sum_{i=\frac{Mm}{r^2}}^{m} C(-x,-y) \right) \\ + \left(\sum_{i=\frac{Mm}{r^2}}^{m} A(x,y) + B(y) + \sum_{i=\frac{Mm}{r^2}}^{m} C(x,y) \right) \end{array} \right\} \frac{Mm}{r^2} \qquad (1.48)$$

We can represent the field equations of the generalized gravitational field in terms of an ordinary linear differential equation of the second-order

$$\left\{ \begin{array}{l} \displaystyle\sum_{i=\frac{Mm}{r^2}}^{m} \left(A\left(x,y,\frac{\partial^2 x}{\partial t^2},\frac{\partial^2 y}{\partial t^2},t\right) + B\left(x,y,\frac{\partial^2 x}{\partial t^2},\frac{\partial^2 y}{\partial t^2},t\right) \right)_{ij} \frac{Mm}{r^2} \\[2em] \displaystyle\sum_{i=\frac{Mm}{r^2}}^{m} C\left(y,\frac{\partial^2 y}{\partial t^2},t\right)_k \frac{Mm}{r^2} + \\[2em] \displaystyle\sum_{i=\frac{Mm}{r^2}}^{m} \left(A\left(x,y,\frac{\partial^2 x}{\partial t^2},\frac{\partial^2 y}{\partial t^2},t\right) + B\left(x,y,\frac{\partial^2 x}{\partial t^2},\frac{\partial^2 y}{\partial t^2},t\right) \right)_{mn} \frac{Mm}{r^2} \end{array} \right\}$$

$$+\left\{ \begin{array}{l} \displaystyle\sum_{i=\frac{Mm}{r^2}}^{m} \left((-)A\left(x,y,\frac{\partial^2 x}{\partial t^2},\frac{\partial^2 y}{\partial t^2},t\right) + (-)B\left(x,y,\frac{\partial^2 x}{\partial t^2},\frac{\partial^2 y}{\partial t^2},t\right) \right)_{ij} \frac{Mm}{r^2} \\[2em] \displaystyle\sum_{i=\frac{Mm}{r^2}}^{m} (-)C\left(y,\frac{\partial^2 y}{\partial t^2},t\right)_k \frac{Mm}{r^2} + \\[2em] \displaystyle\sum_{i=\frac{Mm}{r^2}}^{m} \left((-)A\left(x,y,\frac{\partial^2 x}{\partial t^2},\frac{\partial^2 y}{\partial t^2},t\right) + (-)B\left(x,y,\frac{\partial^2 x}{\partial t^2},\frac{\partial^2 y}{\partial t^2},t\right) \right)_{mn} \frac{Mm}{r^2} \end{array} \right\} \quad (1.49)$$

which written in terms of trigonometric coordinates is

$$\left\{\begin{array}{l} \sum_{i=\frac{Mm}{r^2}}^{m}\left(A\left(r\sin\theta, r\cos\theta, \frac{\partial^2 x}{\partial t^2}, \frac{\partial^2 y}{\partial t^2}, t\right)+B\left(r\sin\theta, r\cos\theta, \frac{\partial^2 x}{\partial t^2}, \frac{\partial^2 y}{\partial t^2}, t\right)\right)_{ij}\frac{Mm}{r^2} \\[2em] \sum_{i=\frac{Mm}{r^2}}^{m} C\left(r\sin\theta, r\cos\theta, \frac{\partial^2 y}{\partial t^2}, t\right)_k \frac{Mm}{r^2}+ \\[2em] \sum_{i=\frac{Mm}{r^2}}^{m}\left(A\left(r\sin\theta, r\cos\theta, \frac{\partial^2 x}{\partial t^2}, \frac{\partial^2 y}{\partial t^2}, t\right)+B\left(r\sin\theta, r\cos\theta, \frac{\partial^2 x}{\partial t^2}, \frac{\partial^2 y}{\partial t^2}, t\right)\right)_{mn}\frac{Mm}{r^2} \end{array}\right\}$$

$$+\left\{\begin{array}{l} \sum_{i=\frac{Mm}{r^2}}^{m}\left((-)A\left(r\sin\theta, r\cos\theta, \frac{\partial^2 x}{\partial t^2}, \frac{\partial^2 y}{\partial t^2}, t\right)+(-)B\left(r\sin\theta, r\cos\theta, \frac{\partial^2 x}{\partial t^2}, \frac{\partial^2 y}{\partial t^2}, t\right)\right)_{ij}\frac{Mm}{r^2} \\[2em] \sum_{i=\frac{Mm}{r^2}}^{m}(-)C\left(r\sin\theta, r\cos\theta, \frac{\partial^2 y}{\partial t^2}, t\right)_k \frac{Mm}{r^2}+ \\[2em] \sum_{i=\frac{Mm}{r^2}}^{m}\left((-)A\left(r\sin\theta, r\cos\theta, \frac{\partial^2 x}{\partial t^2}, \frac{\partial^2 y}{\partial t^2}, t\right)+(-)B\left(r\sin\theta, r\cos\theta, \frac{\partial^2 x}{\partial t^2}, \frac{\partial^2 y}{\partial t^2}, t\right)\right)_{mn}\frac{Mm}{r^2} \end{array}\right\}$$

$$=\left(\sum_{i=\frac{Mm}{r^2}}^{m} A(r\sin\theta, r\cos\theta)+\sum_{i=\frac{Mm}{r^2}}^{m} B(r\sin\theta)+\sum_{i=\frac{Mm}{r^2}}^{m} C(r\sin\theta, r\cos\theta)\right)_{ij}\frac{Mm}{r^2}$$

$$+\sum_{i=\frac{Mm}{r^2}}^{m} C(r\sin\theta)_k \frac{Mm}{r^2}$$

$$+\left(\sum_{i=\frac{Mm}{r^2}}^{m}(-)A(r\sin\theta, r\cos\theta)+\sum_{i=\frac{Mm}{r^2}}^{m}(-)B(r\sin\theta)+\sum_{i=\frac{Mm}{r^2}}^{m}(-)C(r\sin\theta, r\cos\theta)\right)_{ij}\frac{Mm}{r^2} \qquad (1.50)$$

and in terms of linear vector spaces, we will have

$$(\ )\left\{\sum_{i=\frac{Mm}{r^2}}^{m}(F_{xy})_i+\sum_{i=\frac{Mm}{r^2}}^{m}(F_y)_j+\sum_{i=\frac{Mm}{r^2}}^{m}(F_{xy})_k\right\}\frac{Mm}{r^2}+(\pm)\left\{\sum_{i=\frac{Mm}{r^2}}^{m}(F_{xy})_i+\sum_{i=\frac{Mm}{r^2}}^{m}(F_y)_j+\sum_{i=\frac{Mm}{r^2}}^{m}(F_{xy})_k\right\}\frac{Mm}{r^2}$$

when expanded, with respect to the factor of pi we will have

$$\left\{ \sum_{i=\frac{Mm}{r^2}}^{m} \left(F_{xy} \pm \frac{\alpha}{\beta}\pi \right) + \sum_{i=\frac{Mm}{r^2}}^{m} \left(F_{y} \pm \frac{\alpha}{\beta}\pi \right) + \sum_{i=\frac{Mm}{r^2}}^{m} \left(F_{-(xy)} \pm \frac{\alpha}{\beta}\pi \right) \right\} \qquad (1.51)$$

and then expanded into second-order partial derivatives, we will have

$$
\begin{bmatrix}
\sum_{i=\frac{Mm}{r^2}}^{m} \left(\left(\frac{\partial^2 f}{\partial x^2} + \frac{\partial^2 f}{\partial y^2} \right) + \left(\frac{\alpha}{\beta}\pi \right) \right)_{ij} + \sum_{i=\frac{Mm}{r^2}}^{m} \left(\left(\frac{\partial^2 f}{\partial y^2} \right) + \left(\frac{\alpha}{\beta}\pi \right) \right)_{k} + \\
\sum_{i=\frac{Mm}{r^2}}^{m} \left(\left(\frac{\partial^2 f}{\partial x^2} + \frac{\partial^2 f}{\partial y^2} \right) + \left(\frac{\alpha}{\beta}\pi \right) \right)_{mn}
\end{bmatrix} \frac{Mm}{r^2}
$$

$$
+ \begin{bmatrix}
\sum_{i=\frac{Mm}{r^2}}^{m} \left(\left(\frac{\partial^2 f}{\partial x^2} + \frac{\partial^2 f}{\partial y^2} \right) + \left(\frac{\alpha}{\beta}\pi \right) \right)_{(-)ij} + \sum_{i=\frac{Mm}{r^2}}^{m} \left(\left(\frac{\partial^2 f}{\partial y^2} \right) + \left(\frac{\alpha}{\beta}\pi \right) \right)_{(-)k} + \\
\sum_{i=\frac{Mm}{r^2}}^{m} \left(\left(\frac{\partial^2 f}{\partial x^2} + \frac{\partial^2 f}{\partial y^2} \right) + \left(\frac{\alpha}{\beta}\pi \right) \right)_{(-)mn}
\end{bmatrix} \frac{Mm}{r^2}
$$

$$
= \begin{bmatrix}
\sum_{i=\frac{Mm}{r^2}}^{m} \left(\nabla^2 F_{xy} + \left(\frac{\alpha}{\beta}\pi \right) \right) + \sum_{i=\frac{Mm}{r^2}}^{m} \left(\left(\frac{\partial^2 f}{\partial y^2} \right) + \left(\frac{\alpha}{\beta}\pi \right) \right)_{k} + \\
\sum_{i=\frac{Mm}{r^2}}^{m} \left(\nabla^2 F_{mn} + \left(\frac{\alpha}{\beta}\pi \right) \right)
\end{bmatrix} \frac{Mm}{r^2}
$$

$$
+ \begin{bmatrix}
\sum_{i=\frac{Mm}{r^2}}^{m} \left(\nabla^2 F_{xy} + \left(\frac{\alpha}{\beta}\pi \right) \right) + \sum_{i=\frac{Mm}{r^2}}^{m} \left(\left(\frac{\partial^2 f}{\partial y^2} \right) + \left(\frac{\alpha}{\beta}\pi \right) \right)_{k} + \\
\sum_{i=\frac{Mm}{r^2}}^{m} \left(\nabla^2 F_{mn} + \left(\frac{\alpha}{\beta}\pi \right) \right)
\end{bmatrix} \frac{Mm}{r^2} \qquad (1.52)
$$

As has been stated at each point in space, there exists a stress-energy tensor. It defines the energy, momentum, and stress density for that region in space. We also concern ourselves with the conservation of energy-momentum. We can define the stress-energy tensor in terms of components as

$$T_i = T(e_i) = T_j(e_j) = T^i = T(e^i) = T^j(e^j)$$

6

Setting the Riemmanian Tensor to Zero

Introduction

Now given the spacetime metric, the curved and locally Lorentz spacetime metric we can replace it with a new metric as defined not by the curvature of the physical path of the moving particles but by the curvature of the physical motion of the moving particles within range of the gravitational field, used to measure the square of a defined distance or interval between two distinguishable events.

We take a family of paths, each path distinguished from the rest by a select parameter denoted *n*. For the class of paths, we have the covariant derivative and connection coefficients, given as

$$\Gamma^{\alpha}_{\beta\gamma} = \left\langle w^{\alpha}, \nabla_{\gamma} e_{\beta} \right\rangle \tag{1.53}$$

and

$$\nabla_{u} T = T^{\beta}_{\alpha,\gamma} = \partial_{\gamma} T^{\alpha}_{\beta} = T^{\beta}_{\alpha,\gamma} + \Gamma^{\beta}_{\alpha\gamma} T^{u}_{o} - \Gamma^{u}_{\alpha\gamma} T^{\beta}_{u}$$

$$= \left(T^{\beta}_{\alpha,\gamma}, u^{\gamma} \right) e_{\beta} \otimes w^{\alpha} = T^{\beta}_{\alpha,\gamma} u^{\gamma} = T^{\beta}_{\alpha,\gamma} \frac{dx}{d\lambda} \tag{1.54}$$

$$= \frac{DT^{\beta}_{\alpha}}{d\lambda} + \Gamma^{\beta}_{\alpha\gamma} T^{u}_{o} - \Gamma^{u}_{\alpha\gamma} T^{\beta}_{u} \frac{dx^{\gamma}}{d\lambda} = 0$$

Now if the spacetime is flat then the connection coefficients, geodesic equation and curvature tensor vanish, giving us

$$dx^{\alpha} / dt (\gamma) + V = a^{\alpha} + b^{\alpha} \lambda + V \tag{1.55}$$

With

$$d^{2}x^{\beta} / d\lambda^{2} + \Gamma^{\beta}_{uv} \left(dx^{u} / dt \right) \left(dx^{v} / dt \right) = 0$$

$$\Gamma^{\beta}_{uv} = 0 : R^{\beta}_{\gamma uv} = 0 \tag{1.56}$$

While the scalar density of the curvature R can be expressed as

$$\nabla_u \nabla_v + R(\ldots, uvn) = \left(D^2 n^\alpha / d\lambda^2 \right) + R^\alpha_{\beta\gamma\delta} u^\beta n^\gamma v^\sigma = 0 \tag{1.57}$$

where

$$R^\alpha_{\beta x \delta} = \left\langle dx^\alpha, \left[\nabla_\gamma \nabla_\delta \right] e_\beta \right\rangle = \frac{\partial \Gamma^\alpha_{\beta\gamma}}{\partial x^\gamma} - \frac{\partial \Gamma^\eta_{\chi\rho}}{\partial x^\sigma} + \Gamma^\alpha_{u\gamma} \Gamma^u_{\beta\delta} - \Gamma^\alpha_{u\delta} \Gamma^u_{\beta\gamma}$$

$$= R_{uv} = R^\alpha_{\beta x \delta} = \Gamma^\alpha_{uv,a} - \Gamma^\alpha_{ua,v} + \Gamma^\alpha_{u\gamma} \Gamma^u_{\beta\delta} - \Gamma^\alpha_{u\delta} \Gamma^u_{\beta\gamma}$$

$$= -g^{uv} \left(G^\gamma_{uv} G^\sigma_{\lambda\sigma} - G^\gamma_{uv} G^\sigma_{\lambda\sigma} \right) - \partial_v \left(g^{uv} G^\sigma_{u\sigma} - g^{u\sigma} G^v_{u\sigma} \right)$$

$$= D_\lambda G^\lambda_{uv} - D_\lambda G^\lambda_{vu} + G_\lambda G^\lambda_{uv} + G_\lambda G^\lambda_{vu} = \Gamma^\alpha_{uv,\alpha} - \Gamma^\alpha_{ua,v} = \frac{1}{2} \left(h^a_{u,va} + h^a_{v,ua} - h^a_{uv,a} - h_{uv} \right) \tag{1.58}$$

$$= \frac{\partial \Gamma^l_{il}}{\partial x^k} - \frac{\partial \Gamma^l_{ik}}{\partial x^l} + \Gamma^r_{il2} \Gamma^l_{kr} - \Gamma^r_{ik} \Gamma^l_{rl}$$

$$= \frac{\partial \Gamma^r_{ir}}{\partial x^k} - \frac{\partial \Gamma^l_{ik}}{\partial x^r} = \frac{\eta^{rs}}{2} \frac{\partial}{\partial x^k} \left(\frac{\partial h_{si}}{\partial x^k} + \frac{\partial h_{sr}}{\partial x^i} - \frac{\partial h_{ir}}{\partial x^s} \right) - \frac{\eta^{pn}}{2} \frac{\partial}{\partial x^l} \left(\frac{\partial h_{ri}}{\partial x^r} + \frac{\partial h_{ik}}{\partial x^l} - \frac{\partial h_{hu}}{\partial x^m} \right)$$

$$= \frac{\eta^{rs}}{2} \frac{\partial^2 h_{ik}}{\partial x^r \partial x^s} \left(\frac{\partial^2 h}{\partial x^i \partial x^k} + \frac{\partial^2 h^r_i}{\partial x^r \partial x^s} - \frac{\partial^2 h^r_k}{\partial x^r \partial x^s} \right) = \frac{1}{2} \eta_{ik} - \frac{1}{2} \frac{\partial}{\partial x^i} \frac{\partial x^l_k}{\partial x^l} - \frac{1}{2} \frac{\partial}{\partial x^k} \frac{\partial x^v_u}{\partial x^m} = 0$$

In respect to the curvature tensor, as measured by the change in the structure of a vector that is transported around a closed path or route, we will have

$$R(u,v)w = \nabla_u \nabla_v w - \nabla_v \nabla_u - \nabla_{[u,v]} w = 0 \tag{1.59}$$

and

$$\nabla_{e_i} \nabla_{e_k} e_i = \nabla_{e_i} \left[K_{ik} \frac{n}{(n \cdot n)} + \Gamma^m_{ik} e_m \right] = K_{ik,j} \frac{n}{(n \cdot n)} - K_{ik} K^m_j e_m \frac{1}{(n \cdot n)} + \Gamma^m_{ik,j} e_m$$

with

$$R(e_j, e_i) e_i = \left(K_{ikj} - K_{ijk} \right) \frac{n}{(n \cdot n)} + \left[\frac{1}{(n \cdot n)} \left(K_{ij} K^m_k - K_{ik} K^m_j \right) + R^m_{ijk} \right] e_m \tag{1.60}$$

For the linearized theory of gravity, the metric coefficients also vanish as does the overall values of the Riemann tensor, in which the metric perturbation than read

$$R_{uv} = \Gamma^{\alpha}_{uv,\alpha} - \Gamma^{\alpha}_{ua,v} = \frac{1}{2}\left(h^{a}_{u,va} + h^{a}_{v,ua} - h^{a}_{uv,a} - h_{uv}\right) = 0 \qquad (1.61)$$

Which can be written in expanded form

$$R^{uv}(x) - \frac{m^2}{2}\left[g^{uv}(x) - g^{\alpha\beta}g^{uv}\gamma_{\alpha\beta}(x)\right] = 8\pi\left[T^{uv}(x) - \frac{1}{2}g^{uv}T(x)\right] = 0 \qquad (1.62)$$

which for any given gravitating body

$$R = -g^{uv}\left(\Gamma^{\lambda}_{uv}\Gamma^{\sigma}_{\lambda\sigma} - \Gamma^{\lambda}_{u\sigma}\Gamma^{\sigma}_{uv}\right) - \partial_v\left(g^{uv}\Gamma^{\sigma}_{u\sigma} - g^{u\sigma}\Gamma^{u}_{u\sigma}\right)$$
$$= -g^{uv}\left(G^{\lambda}_{uv}G^{\sigma}_{\lambda\sigma} - G^{\lambda}_{u\sigma}G^{\sigma}_{uv}\right) - \partial_v\left(g^{uv}G^{\sigma}_{u\sigma} - g^{u\sigma}G^{v}_{u\sigma}\right)$$

The structure of a flat space-time would resolve to a general Euclidean structure and the equations representing any curvature would be set to

$$B_{uv\sigma} = \frac{\partial}{\partial x_\sigma}\left(\frac{\partial A_u}{\partial x_\sigma} - \{uv,\alpha\}A_\alpha\right) - \{uv,\alpha\}\left(\frac{\partial A_v}{\partial x_\sigma} - \{uv,\alpha\}A_\alpha\right)$$

$$-\{uv,\alpha\}\left(\frac{\partial A_u}{\partial x_\sigma} - \{uv,\alpha\}A_\alpha\right) = \frac{\partial^2 A_u}{\partial x_\sigma \partial x_v} - \{uv,\alpha\}\frac{\partial A_u}{\partial x_\sigma} - \{uv,\alpha\}\frac{\partial A_v}{\partial x_\sigma} - \{uv,\alpha\}\frac{\partial A_u}{\partial x_\sigma}$$

$$+\{uv,\alpha\}A_e + \{uv,\alpha\}\{uv,\alpha\}A_e - A_\sigma\{uv,\alpha\}\frac{\partial A_u}{\partial x_\sigma} = 0$$

$$A_{auv} - A_{bvu} = A_e\left(\{uv,\alpha\}\{uv,\alpha\} - \{uv,\alpha\}\frac{\partial A_u}{\partial x_\sigma} - \{uv,\alpha\}\{uv,\alpha\} + \{uv,\alpha\}\frac{\partial A_u}{\partial x_\sigma}\right)$$

$$= A_e B^{e}_{uva} = 0$$

$$B^{e}_{uva} = \{uv,\alpha\}\{uv,\alpha\} - \{uv,\alpha\}\{uv,\alpha\} + \{uv,\alpha\}\frac{\partial A_u}{\partial x_\sigma} - \{uv,\alpha\}\frac{\partial A_u}{\partial x_\sigma} = 0$$

$$B^{e}_{uva} = \{uv,\alpha\}\{uv,\alpha\} - \{uv,\alpha\}\{uv,\alpha\} + \{uv,\alpha\}\frac{\partial A_u}{\partial x_\sigma} - \{uv,\alpha\}\frac{\partial A_u}{\partial x_\sigma} = 0$$

Which reduces to

$$R = -g^{uv}\left(\Gamma^{\lambda}_{uv}\Gamma^{\sigma}_{\lambda\sigma} - \Gamma^{\lambda}_{u\sigma}\Gamma^{\sigma}_{v\lambda}\right) - \partial_v\left(g^{uv}\Gamma^{\sigma}_{u\sigma} - g^{u\sigma}\Gamma^{v}_{u\sigma}\right)$$
$$= -g^{uv}\left(G^{\lambda}_{uv}G^{\sigma}_{\lambda\sigma} - G^{\lambda}_{u\sigma}G^{\sigma}_{v\lambda}\right) - \partial_v\left(g^{uv}G^{\sigma}_{u\sigma} - g^{u\sigma}G^{v}_{u\sigma}\right) = 0$$

The variation the Riemann tensor is given as

$$\delta_\xi R_{uv} = -R_{u\sigma}D_v\xi^\sigma - R_{u\sigma}D_v\xi^\sigma - R_{u\sigma}D_v\xi^\sigma = 0$$

Where *R* is previously defined as

$$R_{uv} = D_\lambda G_{uv}^\lambda - D_\lambda G_{uv}^\lambda + -G_\lambda G_{uv}^\lambda - G_\lambda G_{uv}^\lambda = 0$$

For the linearized theory of gravity, the metric coefficients in the metric perturbation reads

$$R_{uv} = \Gamma_{uv,\alpha}^\alpha - \Gamma_{u\alpha,v}^\alpha = \frac{1}{2}\left(h_{u,v\alpha}^\alpha + h_{v,u\alpha}^\alpha - h_{uv,\alpha}^\alpha - h_{uv}\right) = 0$$

and in respect to the curvature tensor, as measured by the change in structure of a vector transported around a closed route or path, we have

$$R(u,v)w = \nabla_u\nabla_v w - \nabla_v\nabla_u w - \nabla_{[u,v]}w = 0$$

With

$$R(e_j,e_i)e_i = \left(K_{ikj} - K_{ijk}\right)\frac{n}{(n\cdot n)} + \left[(n\cdot n)^{-1}\left(K_{ij}K_k^m - K_{ik}K_j^m\right) + R_{ijk}^m\right]e_m = 0$$

and

$$R_{ijk}^m = R_{ijk}^m + (n\cdot n)^{-1}\left(K_{ij}K_k^m - K_{ik}K_j^m\right) = 0$$

$$R_{ijk}^n = (n\cdot n)^{-1}R_{nijk} = -(n\cdot n)^{-1}\left(K_{ijk} - K_{ikj}\right) = 0$$

$$R_{ink}^n = (n\cdot n)^{-1}\left(\frac{\partial K_{ik}}{\partial n} + K_{im}K_k^m\right) = 0$$

by introducing a new tensor, we obtain the equations

$$\nabla_i e_j = K_{ij}\frac{n}{n\cdot n} + \left(\Gamma_{ji}^u e_u\right) = \nabla_i A = A_i^j e_j + K_{ij}A^j\frac{n}{n\cdot n} \tag{1.63}$$

Where

$$(dn)_i = n_{ik}dx^k = \left(\frac{\partial n}{\partial x^k} + \Gamma_{ik}^\sigma n_\sigma\right)dx^k = K_{ik}dx^k \tag{1.64}$$

And

$$\nabla_i n = -K(e_i) = -K_i^j e^j \rightarrow K_{im} = K_i^j \left(e_j \cdot e_m \right) = -e_m \cdot \nabla_i n$$

$$= n \cdot \nabla_i e_m = \left(n \cdot e_0 \right) \Gamma_{mi}^0 = n \cdot \nabla_m e_i = K_{mi} \tag{1.65}$$

with

$$K_{ik} = -n_{lk} = -N\Gamma_{ik}^0 = -N\left[g^{00}\Gamma_{0ik} + g^{0p}\Gamma_{pik} \right] = (1/N)\left[\Gamma_{0ik} - N^p\Gamma_{pik} \right]$$

$$= \frac{1}{2N}\left[\frac{\partial N_i}{\partial x^k} + \frac{\partial N_k}{\partial x^i} - \frac{\partial g_{ik}}{\partial t} - 2N^p\Gamma_{pik} \right] = \frac{1}{2N}\left[N_{ik} + N_{kl} - \frac{\partial g_{ik}}{\partial t} \right] \tag{1.66}$$

and in respect to the curvature tensor, as measured by the change in structure of a vector transported around a closed route or path, we have

$$R(u,v)w = \nabla_u \nabla_v w - \nabla_v \nabla_u w - \nabla_{[u,v]} w = 0 \tag{1.67}$$

And

$$\nabla_{e_j} \nabla_{e_k} e_i = \nabla_{e_j}\left[K_{ik} \frac{n}{n \cdot n} + \left(\Gamma_{ik}^m e_m \right) \right] = K_{ik,j} \frac{n}{(n \cdot n)} - K_{ik}K_j^m e_m \frac{1}{(n \cdot n)} + \left(\Gamma_{ik,j}^m e_m \right) \tag{1.68}$$

With

$$R\left(e_i \cdot e_j \right) e_i = \left(K_{ikj} - K_{ijk} \right) \frac{n}{(n \cdot n)} + \left[(n \cdot n)^{-1}\left(K_{ij}K_k^m - K_{ik}K_j^m \right) + R_{ijk}^m \right] e_m = 0 \tag{1.69}$$

and

$$R_{ijk}^m = R_{ijk}^m + (n \cdot n)^{-1}\left(K_{ij}K_k^m - K_{ik}K_j^m \right) = 0$$

$$R_{ijk}^n = (n \cdot n)^{-1} R_{nijk} = -(n \cdot n)^{-1}\left(K_{ijk} - K_{ikj} \right) \tag{1.70}$$

$$R_{ijk}^m = (n \cdot n)^{-1}\left(\frac{\partial K_{ik}}{\partial n} + K_{im}K_k^m \right) = 0$$

7

G.R.F.F. Approximation of the Riemann Curvature Tensor

Introduction

Here we find the mathematical replacement for the original Riemannian Curvature Tensor, denoted as the G.R.F.F. special linear vector spaces, or Gravitational Radiation Flux Force Field, without compromising its original form.

Replacing the original Riemannian Curvature Tensor

Here we have the simple notation of $(+)$ when adding normal vectors or forces, and $(\oplus)$ when adding special linear vector spaces.

We begin the fist set of equations

$$R_{00} = R_{00} + R_{00} + \ldots = \Gamma^i_{00,i(2)} = \Gamma^i_{00,i(2)} - \Gamma^i_{00} + \Gamma^i_{00}\Gamma^j_{ij} - \Gamma^0_{0i}\Gamma^i_{00}$$

$$= \frac{1}{2}\Delta 2\left(F_{(uv)_\alpha} \oplus F_{(uv)_\beta}\right)_2 = \frac{1}{2}\Delta 2\left(F_{(uv)_\alpha} \oplus F_{(uv)_\beta}\right)_4$$

$$-g_{0i,0j} + \left(F_{(uv)_\alpha} \oplus F_{(uv)_\beta}\right)\delta_{ij}2\left(F_{(uv)_\alpha} \oplus F_{(uv)_\beta}\right)$$

$$+\left(F_{(uv)_\alpha} \oplus F_{(uv)_\beta}\right)\delta_{ij}2\left(F_{(uv)_\alpha} \oplus F_{(uv)_\beta}\right)2\left(F_{(uv)_\alpha} \oplus F_{(uv)_\beta}\right)\delta_{ij}$$

$$+\left(F_{(uv)_\alpha} \oplus F_{(uv)_\beta}\right)\delta_{ij,j}2\left(F_{(uv)_\alpha} \oplus F_{(uv)_\beta}\right)$$

$$-\frac{1}{2}\left(F_{(uv)_\alpha} \oplus F_{(uv)_\beta}\right)_j 2\left(F_{(uv)_\alpha} \oplus F_{(uv)_\beta}\right)\delta_{jj,i} - \frac{1}{2}\left(F_{(uv)_\alpha} \oplus F_{(uv)_\beta}\right)_{,i}2\left(F_{(uv)_\alpha} \oplus F_{(uv)_\beta}\right)_{,j}$$

$$= \frac{1}{2}\Delta 2\left(F_{(uv)_\alpha} \oplus F_{(uv)_\beta}\right) + 2\left(F_{(uv)_\alpha} \oplus F_{(uv)_\beta}\right)\Delta\left(F_{(uv)_\alpha} \oplus F_{(uv)_\beta}\right)$$

$$-2\left(\nabla\left(F_{(uv)_\alpha} \oplus F_{(uv)_\beta}\right)\right)^2 \tag{1.71}$$

And

$$R_{ij} = R_{ii} + R_{jj} + \ldots = \Gamma^k_{jk,k} - \Gamma^0_{i0,j} - \Gamma^k_{ik,j}$$

$$= -\left(F_{(uv)_\alpha} \oplus F_{(uv)_\beta}\right)2\left(F_{(uv)_\alpha} \oplus F_{(uv)_\beta}\right)\delta_{ij} + \left(F_{(uv)_\alpha} \oplus F_{(uv)_\beta}\right)\delta_{kk}2\left(F_{(uv)_\alpha} \oplus F_{(uv)_\beta}\right)\delta_{ij}$$

$$-\left(F_{(uv)_\alpha} \oplus F_{(uv)_\beta}\right)\delta_{ik}2\left(F_{(uv)_\alpha} \oplus F_{(uv)_\beta}\right)\delta_{kj} - \left(F_{(uv)_\alpha} \oplus F_{(uv)_\beta}\right)\delta_{kj}2\left(F_{(uv)_\alpha} \oplus F_{(uv)_\beta}\right)\delta_{ki}$$

$$+\frac{1}{2}\Delta 2\left(F_{(uv)_\alpha} \oplus F_{(uv)_\beta}\right)\delta_{ij} \tag{1.72}$$

the simplified tensor can be given as

$$R_{ij} = \frac{1}{2}\Delta 2\left(F_{(uv)_\alpha} \oplus F_{(uv)_\beta}\right)\delta_{ij}$$

$$R_{00} = \frac{1}{2}\Delta 2\left(F_{(uv)_\alpha} \oplus F_{(uv)_\beta}\right) + 2\left(F_{(uv)_\alpha} \oplus F_{(uv)_\beta}\right)\Delta - 2\left(\nabla\left(F_{(uv)_\alpha} \oplus F_{(uv)_\beta}\right)\right) \tag{1.73}$$

The density of the mass-energy as measured by any observer in motion is replaced with the mass-gravity tensor

$$T_{00} = -T^0_0 = T^{00} = T(e_0, e_0) \to \Gamma(v_0, m_0) \tag{1.74}$$

The metric tensor for the gravitational field can be given as

$$\Phi \to F\left(V_{ijk}, V_{-(ijk)}\right)$$

$$= \left(\frac{-F_{ijk}}{\left(r_{xyz}^2\right)^{3/2}} i + \frac{-F_{ijk}}{\left(r_{xyz}^2\right)^{3/2}} j + \frac{-F_{ijk}}{\left(r_{xyz}^2\right)^{3/2}} k\right) \otimes \left(\frac{F_{ijk}}{\left(r_{xyz}^2\right)^{3/2}} i + \frac{F_{ijk}}{\left(r_{xyz}^2\right)^{3/2}} j + \frac{F_{ijk}}{\left(r_{xyz}^2\right)^{3/2}} k\right)$$

$$= \left(\frac{-F_{ijk}}{\left(x^2+y^2+z^2\right)^{3/2}} i + \frac{-F_{ijk}}{\left(x^2+y^2+z^2\right)^{3/2}} j + \frac{-F_{ijk}}{\left(x^2+y^2+z^2\right)^{3/2}} k\right) \tag{1.75}$$

$$\otimes \left(\frac{F_{ijk}}{\left(x^2+y^2+z^2\right)^{3/2}} i + \frac{F_{ijk}}{\left(x^2+y^2+z^2\right)^{3/2}} j + \frac{F_{ijk}}{\left(x^2+y^2+z^2\right)^{3/2}} k\right)$$

When considering an arbitrary mass, denoted by the generalized tensor M, our metric equations become

$$F_T = \left(\Gamma_M, \Gamma_{uv}\right) = \left(\Gamma_M, \Gamma_u, \Gamma_v\right) = F_{Muv} M \otimes \left(F_{(uv)_\alpha} \oplus F_{(uv)_\beta}\right) \tag{1.76}$$

where this can be given in components as

$$\Gamma_{uv} = V_u \otimes V_v = \Gamma\left(V_u, V_v\right) \tag{1.77}$$

than the two divergent and convergent vector fields, as a dual-based symmetric second order tensor, can be given in terms of commutators as

$$\left[F_u, F_v\right] = \left[\partial_u, \partial_v\right] = \partial_v \partial_u - \partial_u \partial_v \tag{1.78}$$

in spherical coordinates, this can be given as

$$\nabla \cdot F_{ijk} = \frac{\partial F_u}{\partial x} + \frac{\partial F_v}{\partial y} + \frac{\partial F_w}{\partial z} = \frac{1}{r^2}\frac{\partial}{\partial r}\left(r^2 F_r\right) + \frac{1}{r\sin\theta}\frac{\partial}{\partial\theta}\left(\sin\theta F_\theta\right) + \frac{1}{r\sin\theta}\frac{\partial F_\phi}{\partial\phi} \tag{1.79}$$

which in terms of the mass-gravity tensor is given in spherical coordinates

$$\nabla \cdot F_{ijk} \otimes \nabla \cdot F_{-(ijk)} = \left(\frac{\partial F_u}{\partial x} + \frac{\partial F_v}{\partial y} + \frac{\partial F_w}{\partial z} \right) \otimes (-) \left(\frac{\partial F_u}{\partial x} + \frac{\partial F_v}{\partial y} + \frac{\partial F_w}{\partial z} \right)$$

$$= \left(\frac{1}{r^2} \frac{\partial}{\partial r} \left(r^2 (-) F_r \right) + \frac{1}{r \sin \theta} \frac{\partial}{\partial \theta} \left(\sin \theta (-) F_\theta \right) + \frac{1}{r \sin \theta} \frac{\partial (-) F_\phi}{\partial \phi} \right) \qquad (1.80)$$

$$\otimes \left(= \frac{1}{r^2} \frac{\partial}{\partial r} \left(r^2 F_r \right) + \frac{1}{r \sin \theta} \frac{\partial}{\partial \theta} \left(\sin \theta F_\theta \right) + \frac{1}{r \sin \theta} \frac{\partial F_\phi}{\partial \phi} \right)$$

Giving us the mass-gravity tensor

$$T \left\langle dx^u / dt (\gamma) + V \right\rangle = VT \left\langle dx^u / dt (\gamma) + V \right\rangle = VT^{u0} \qquad (1.81)$$

Then for that region of space conservation of momentum-energy must be frame-dependent giving us

$$p^u = \left\langle dx^u / dt (\gamma) + V, p(\gamma) + V \right\rangle \qquad (1.82)$$

Newtonian world lines read as free-falling particles or inertial particles

$$\frac{d^2 x}{dt^2} + \frac{\partial \Phi}{\partial x^j} \rightarrow \frac{d^2 x}{dt^2} + \frac{\partial}{\partial x^j} \left(F_{(uv)_\alpha} \oplus F_{(uv)_\beta} \right) \qquad (1.83)$$

where

$$L_g = \left(M_i, \Gamma_{uv} \right) = \left(M_i, V_{ij}, V_{kl} \right) \qquad (1.84)$$

8

Reformulation of the original Lagrangian and Lagrangian density

Introduction

We will take the original Lagrangian and Lagrangian density and reformulate these equations in terms of the G.R.F.F. potential.

The Lagrange function *L*, as a scalar function, or simply the Langrangian density, can be rewritten in terms of the original Newtonian density

$$L(x,t) = T_{\alpha\beta} = -p(x,t)U(x,t) \tag{1.85}$$

replaced with the G.R.F.F. potential

$$L(x,t) = T(v) - V(t) = \frac{1}{2}m|dx(t)|^2 - m[F_{uvw}](x(t),t) \tag{1.86}$$

where *U* is the newly defined divergent vector field within 3-dimensions.

This can be written in terms of the newly developed *G.R.F.F.* potential Lagrangian

$$L(t) = T(v) - V(t) = \frac{1}{2}m|dx(t)|^2 - m[F_{uvw}](x(t),t) \tag{1.87}$$

then by varying the x factor in the integral, being equivalent to the Euler-Lagrange differential equation, we will have

$$\delta \int L(t)\,dt = \int \delta L(t)\,dt = \int \left(mdx(t)^2 \cdot \delta dx(t) - m\nabla[F_{uvw}](x(t),t) \cdot x(t) \right) dt \tag{1.88}$$

The Lagrangian density for newly developed gravitational or *G.R.F.F.* potential will be given as

$$L_g = L(x,t) = \rho(x,t)\left[F_{uvw}\right](x,t)$$
$$-\left\{\frac{1}{\nabla^2\left[F_{uv_{(\alpha)}}\right](x,t)}\left(\nabla\left[F_{uv_{(\alpha)}}\right](x,t)\right)^2 \oplus \frac{1}{\nabla^2\left[F_{uv_{(\beta)}}\right](x,t)}\left(\nabla\left[F_{uv_{(\beta)}}\right](x,t)\right)^2\right\} \quad (1.89)$$

variation of the integral will give us

$$\delta L_g = \delta L(x,t) = \rho(x,t)\delta\left[F_{uvw}\right](x,t)$$
$$-\left\{\begin{array}{l}\dfrac{2}{\nabla^2\left[F_{uv_{(\alpha)}}\right](x,t)}\left(\nabla\left[F_{uv_{(\alpha)}}\right](x,t)\right)\left(\nabla\left[F_{uv_{(\beta)}}\right](x,t)\right) \\[4mm] \oplus \dfrac{2}{\nabla^2\left[F_{uv_{(\beta)}}\right](x,t)}\left(\nabla\left[F_{uv_{(\beta)}}\right](x,t)\right)\left(\nabla\delta\left[F_{uv_{(\beta)}}\right](x,t)\right)\end{array}\right\} \quad (1.90)$$

Where the density of the gravity-mass tensor is written as (the

source of the gravitational field) in proportion to the mass of the stellar body, which is proportional to its angular velocity as defined by its internal density times its size as determined by the diameter of its body, giving us

$$T_{\alpha\beta} = -\frac{2\dfrac{\delta L}{\delta T_{uv}}}{M} \quad (1.91)$$
$$\frac{\delta L}{\delta T_{uv}} = \frac{\delta L}{\delta T_{ab}} - \partial_\varsigma \frac{\delta L}{\delta T_{ab}}$$

Here *L* is defined above in equation (1.81).

Invariant volume element

The invariant volume element that constitutes the source of the *G.R.F.F.* tensor, which we write as

$$\sqrt{-\gamma}d^4x = \sqrt{-g}d^4x = g = \det(g_{uv}) \rightarrow \Gamma_M \tag{1.92}$$

Which summed over, gives us the total area of the gravitational source

$$\sum \sqrt{-\gamma}d^4x = \sum \sqrt{-g}d^4x = \sum g = \det(g_{uv}) \rightarrow \Gamma_M \tag{1.93}$$

In terms of the unconserved density of the charged vector current written as the source of the electromagnetic field, is defined by the density of the new vector potential, similar to the *G.R.F.F.* potential,

$$A^v = \left(\phi, \left\{A^v(\gamma) + V\right\}\right)$$
$$\gamma^{\alpha\beta} D_\alpha D_\beta \left\{A^v(\gamma) + V\right\} + u^2 \left\{A^v(\gamma) + V\right\} = 4\pi J \tag{1.94}$$

Which varies corresponding transitions between frames, unprimed to primed, giving us

$$dA^v = \left(\phi, \left\{A^v(\gamma) + V\right\}\right)$$
$$\gamma^{\alpha\beta} D_\alpha D_\beta \left\{A^v(\gamma) + V\right\} + u^2 \left\{A^v(\gamma) + V\right\} = 4\pi J \tag{1.95}$$

The tensor of the gravitational field can then be written in like form giving us its vanishing components

$$\psi^{\alpha\beta} D_\alpha D_\beta \phi^{uv} + u^2 \phi_{\alpha\beta} = \lambda T = 0 \tag{1.96}$$

such that

$$T_{\alpha\beta} = T_{\alpha\beta g} + T_{\alpha\beta M} \rightarrow F_{\alpha\beta\sigma} + T_M = F_T \tag{1.97}$$

Which corresponds to the combined motion of all nearby test particles, assuming they take the form of geodesics, then we have

$$\frac{\sum\left(D_\alpha D_\beta \phi^{\alpha\beta} + \begin{Bmatrix} \gamma \\ \alpha\beta \end{Bmatrix} u^2 \phi^{\alpha\beta}\right)}{\sum \lambda T_{uv}}$$

$$\Gamma_{\alpha\beta} = \Gamma_M^{\alpha\beta}$$

(1.98)

The Lagrangian of matter

The scalar density of the Lagrangian of matter may be written as, vanishing components

$$L = L_g\left(T_{\alpha\beta}, g^{uv}\right) + L_M\left(T_{\alpha\beta}, g^{uv}, \phi_A\right) \rightarrow L_T = L_g\left(M_{uv}, F^{\alpha\beta}\right) + L_M\left(M_{uv}, F^{\alpha\beta}, A_\phi\right)$$

(1.99)

Or in terms of the metric, we have

$$L = L_g\left(T_{\alpha\beta}, g^{uv}\right) + L_M\left(T_{\alpha\beta}, g^{uv}, \phi_A\right) \rightarrow L_T = L_g\left(M_{uv}, F^{\alpha\beta}\right) + L_M\left(M_{uv}, F^{\alpha\beta}, A_\phi\right)$$

(1.100)

If we set for

$$L = L_g\left(T_{\alpha\beta}, g^{uv}\right) + L_M\left(T_{\alpha\beta}, g^{uv}\right) = 0$$

(1.101)

Which after neglecting the curvature of empty space can be set to

$$L = L_M\left(T_{\alpha\beta}, \phi^{\alpha\beta}, \phi_A\right) = 0 \rightarrow L_M\left(\Gamma_{uv}, \Gamma^{\alpha\beta}, \Gamma_M\right)$$
$$L = L_M\left(T_{\alpha\beta}, g^{uv}, \phi_A\right) = 0 \rightarrow L_M\left(\Gamma_{uv}, \Gamma^{\alpha\beta}, \Gamma_M\right)$$

(1.102)

We also obtain the formula for the variation of the Lagrangian, giving us

$$\frac{\delta L}{\delta T_{\alpha\beta}} = \frac{\delta^* L}{\delta T_{\alpha\beta}} + \frac{\delta^* L}{\delta g^{\alpha\beta}}\frac{\partial g^{\alpha\beta}}{\delta T_{\alpha\beta}} = -2\frac{\delta^* L}{\delta g^{\alpha\beta}} \cdot \frac{\partial g^{\alpha\beta}}{\delta \gamma_{uv}} - 2\frac{\delta^* L}{\delta T_{\alpha\beta}}\frac{1}{16\pi} = T_{\alpha\beta}$$

$$\rightarrow \frac{\delta L}{\delta T_{uv}} = \frac{\delta^* L}{\delta F_{\alpha\beta}} + \frac{\delta^* L}{\delta F^{\alpha\beta}}\frac{\partial F^{\alpha\beta}}{\delta T_{\alpha\beta}} = -2\frac{\delta^* L}{\delta F^{\alpha\beta}} \cdot \frac{\partial F^{\alpha\beta}}{\delta \gamma_{uv}} - 2\frac{\delta^* L}{\delta T_{\alpha\beta}}$$

(1.103)

where

$$\left(F_{uv_{(\alpha)}} \oplus F_{uv_{(\beta)}}\right)^2 = \left\langle \omega^\alpha, \omega^\beta \right\rangle$$

(1.104)

If the density of the Lagrangian of matter assumes the form

$$L_M\left(g^{uv}, \phi_A\right) \to L_M\left(\Gamma^{uv}, \Gamma_M\right) \tag{1.105}$$

Which is invariant under the action

$$S_M = \int L_M\left(g^{uv}, \phi_A\right)d^4x \to \int L_M\left(\Gamma^{uv}, \Gamma_M\right)d^3x \tag{1.106}$$

And under the transformation of coordinates

$$\frac{dx'}{dt} = \frac{dx}{dt}(\gamma) + V + \xi\left(\frac{dx}{dt}(\gamma)\right) + V \tag{1.107}$$

The field functions then vary under a set of gauge transformations, giving us the lie variations, of whom satisfy the given conditions

$$\left[\delta_{\xi_1}, \delta_{\xi_1}\right](\cdot) = \delta_{\xi_1}(\cdot)\left[\delta_{\xi_1}\left[\delta_{\xi_1}, \delta_{\xi_1}\right]\right] + \left[\delta_{\xi_2}\left[\delta_{\xi_2}, \delta_{\xi_2}\right]\right] + \left[\delta_{\xi_3}\left[\delta_{\xi_3}, \delta_{\xi_3}\right]\right] \tag{1.108}$$

The variation of the action can then be written in the form

$$\delta_C S_M = \int_\Omega \left[\delta L_M(x) + \partial_\alpha\left(\xi^\alpha L_M(x)\right)\right]d^3x \tag{1.109}$$

And

$$\delta_C S_M = \int_\Omega \left[\delta L_M(x) + \partial_\alpha\left(\xi^\alpha L_M(x)\right)\right]d^3x = 0 \tag{1.110}$$

Where

$$\begin{aligned}
\delta L_M(x) &= \frac{\partial L_M}{\partial g^{uv}}\delta g^{uv} + \frac{\partial L_M}{\partial\left(\partial_\alpha g^{uv}\right)}\delta\left(\partial_\alpha g^{uv}\right) + \frac{\partial L_M}{\partial\phi_A}\delta\phi_A + \frac{\partial L_M}{\partial\left(\partial_\alpha\phi_A\right)}\delta\left(\partial_\alpha\phi_A\right) \\
&= -\partial_\alpha\left[\xi^\alpha(x)L_M(x)\right] \\
&\to \frac{\partial L_M}{\partial g^{uv}}\delta\Gamma^{uv} + \frac{\partial L_M}{\partial\left(\partial_\alpha\Gamma^{uv}\right)}\delta\left(\partial_\alpha\Gamma^{uv}\right) + \frac{\partial L_M}{\partial\Gamma_M}\delta\Gamma_M + \frac{\partial L_M}{\partial\left(\partial_\alpha\Gamma_M\right)}\delta\left(\partial_\alpha\Gamma_M\right) \\
&= -\partial_\alpha\left[\xi^\alpha(x)L_M(x)\right]
\end{aligned} \tag{1.111}$$

also

$$\delta L_M \left(g^{uv}(x) \right) = \frac{\partial L_M}{\partial g^{uv}} \delta g^{uv} + \frac{\partial L_M}{\partial \left(\partial_\alpha g^{uv} \right)} \delta \left(\partial_\alpha g^{uv} \right) + \frac{L_M}{\partial \left(\partial_\alpha \partial_\beta g^{uv} \right)} \delta \left(\partial_\alpha \partial_\beta g^{uv} \right)$$

$$= -\partial_\alpha \left[\xi^\alpha (x) L_M \left(g^{uv}(x) \right) \right]$$

$$\rightarrow \frac{\partial L_M}{\partial \Gamma^{uv}} \delta \Gamma^{uv} + \frac{\partial L_M}{\partial \left(\partial_\alpha \Gamma^{uv} \right)} \delta \left(\partial_\alpha \Gamma^{uv} \right) + \frac{L_M}{\partial \left(\partial_\alpha \partial_\beta \Gamma^{uv} \right)} \delta \left(\partial_\alpha \partial_\beta \Gamma^{uv} \right)$$

$$= -\partial_\alpha \left[\xi^\alpha (x) L_M \left(\Gamma^{uv}(x) \right) \right]$$

(1.112)

The equation of the gravitational field can then be written in the form

$$\frac{\delta L_g}{\delta g^{uv}} = \lambda_1 R_{uv} + \frac{1}{2} \lambda_3 \gamma_{uv} + \lambda_3 \gamma_{uv} = \frac{\delta L_M}{\delta g^{uv}} - \partial_\sigma \left(\frac{\partial L}{\partial \left(\partial_\sigma g^{uv} \right)} \right)$$

$$\rightarrow L_T \left(\Gamma_M, \Gamma_{uv}; \Gamma_{\alpha\beta} \right) = \left(\Gamma_{uv} \oplus \Gamma_{\alpha\beta} \right) \otimes \Gamma_M$$

(1.113)

$$= \frac{\delta L_M}{\delta \left(\Gamma_{uv} \oplus \Gamma_{\alpha\beta} \right)} - \partial_\sigma \left(\frac{\partial L}{\partial \left(\partial_\sigma \left(\Gamma_{uv} \oplus \Gamma_{\alpha\beta} \right) \right)} \right)$$

The density of the Lagrangian can then be written as

$$L_g = \lambda_2 \sqrt{-g} + \lambda_3 \gamma_{uv} g^{uv} + \lambda_4 \sqrt{-\lambda}$$

$$\rightarrow L(x,t) = -\Gamma_M (x,t) \Gamma_g (x,t) \oplus \Gamma_M (x,t) \Gamma_g (x,t)$$

(1.114)

Giving us the transformation equations

$$g'_{\alpha\beta} = g_{\alpha\beta}(x) + \delta_\xi g_{\alpha\beta} + \xi^u (x) D_\alpha g_{\alpha\beta}(x)$$

$$\phi'_{\alpha\beta} = \phi_{\alpha\beta}(x) + \delta_\xi \phi_{\alpha\beta} + \xi^u (x) D_\alpha \phi_{\alpha\beta}(x)$$

$$\delta_\xi g_{\alpha\beta}(x') = g_{\alpha\beta} D_u \xi^\beta (x) + g_{\alpha\beta} D_\beta \xi^u (x) - D_\alpha \left(\xi^\beta g_{\alpha\beta} \right)$$

$$\phi_A (x) = -\xi^\alpha (x) D_\alpha \phi_A (x) + F_{A;B}^{\beta;\alpha} \phi_\beta D_\alpha \varepsilon^\alpha (x)$$

$$\delta_\varepsilon \phi_{\alpha\beta}(x') = g_{\alpha u} D_\alpha \varepsilon^v (x) + g_{\beta u} D_\alpha \varepsilon^u (x) - D_\alpha a \left(\varepsilon^\alpha g_{\alpha\beta} \right)$$

$$\delta_\xi \phi_{\alpha\beta}(x') = \phi_{\alpha u} D_u \xi^v (x) + \phi_{\beta u} D_\alpha \xi^u (x) - D_\alpha a \left(\xi^\alpha \phi_{\alpha\beta} \right)$$

(1.115)

given then as

$$\Gamma'_{\alpha\beta}(x') = \Gamma_{\alpha\beta}(x) + \delta_{\xi}\Gamma_{\alpha\beta}(x) + \xi^{u}(x)D_{\alpha}\Gamma_{\alpha\beta}(x)$$

$$\delta_{\xi}\Gamma_{\alpha\beta}(x') = \Gamma_{\alpha\beta}D_{u}\xi^{\beta}(x) + \Gamma_{\alpha\beta}D_{\beta}\xi^{u}(x) - D_{\alpha}\left(\xi^{\beta}\Gamma^{\alpha\beta}\right) \qquad (1.116)$$

$$\delta_{\varepsilon}\phi_{\alpha\beta}(x) = \Gamma_{\alpha u}D_{\alpha}\varepsilon^{v}(x) + \Gamma_{\beta u}D_{\alpha}\varepsilon^{u}(x) - D_{\alpha}a\left(\varepsilon^{\beta}\Gamma_{\alpha\beta}\right)$$

Where

$$G^{\lambda}_{uv} = \frac{1}{2}g^{\gamma\sigma}\left(D_{u}\,g^{\sigma v} + D_{v}\,g^{\sigma v} - D_{\sigma}\,g^{uv}\right) \qquad (1.117)$$

Where the curvature of the Riemann tensor vanishes, the function of the deviation of all nearby geodesics is placed in place of the Riemann tensor, giving us

$$\sum_{i=1}^{n}\left(\frac{d^2x_{\alpha}}{ds^2} + \{uv,\alpha\}\frac{dx_u}{ds}\frac{dx_v}{ds}\right) = \frac{d^2x_{\alpha}}{ds^2} + \{11,1\}\frac{dx_u}{ds}\frac{dx_v}{ds}$$

$$+ \frac{d^2x_{\alpha}}{ds^2} + \{12,1\}\frac{dx_u}{ds}\frac{dx_v}{ds} + \frac{d^2x_{\alpha}}{ds^2} + \{12,2\}\frac{dx_u}{ds}\frac{dx_v}{ds} + \frac{d^2x_{\alpha}}{ds^2} + \{uv,\alpha\}\frac{dx_u}{ds}\frac{dx_v}{ds}$$

the metric is in terms of an expansion, is given as

$$2F_{\alpha\beta\gamma} = 1 + 2F_{\alpha\beta\gamma(2)} + 2F_{\alpha\beta\gamma(4)} + 2F_{\alpha\beta\gamma(6)} \qquad (1.118)$$

The equation of the gravitational field can then be written in the form

$$\frac{\delta L_g}{\delta g^{uv}} = \lambda_1 R_{uv} + \frac{1}{2}\lambda_2 g^{uv} + \lambda_3 \gamma_{uv} = \frac{\delta L_g}{\delta g^{uv}} - \partial_{\sigma}\left(\frac{\delta L}{\partial\left(\partial_{\sigma}g^{uv}\right)}\right)$$

$$\rightarrow L_T\left(\Gamma_M,\Gamma_{uv};\Gamma_{\alpha\beta}\right) = \left(\Gamma_{uv}\oplus\Gamma_{\alpha\beta}\right)\otimes\Gamma_M \qquad (1.119)$$

$$= \frac{\delta L_g}{\delta\left(\Gamma_{uv}\oplus\Gamma_{\alpha\beta}\right)^{uv}} - \partial_{\sigma}\left(\frac{\delta L}{\partial\left(\partial_{\sigma}\left(\Gamma_{uv}\oplus\Gamma_{\alpha\beta}\right)^{uv}\right)}\right)$$

The total density of the Lagrangian of matter

$$L = L_M\left(g^{uv},\phi_A\right) \rightarrow L_T\left(\Gamma_M,\Gamma_{uv};\Gamma_{\alpha\beta}\right) \qquad (1.120)$$

The variation of the action is given as

$$\delta_C S_M = \delta_C \int L_M \left(g^{uv}, \phi_A \right) d^4 x \to \delta_C \int L_T \left(\Gamma_M, \Gamma_{uv}; \Gamma_{\alpha\beta} \right) d^3 x \qquad (1.121)$$

where

$$\delta L_g = \left\{ \rho(x,t) \delta F_{\alpha\beta\sigma}(x,t) - \frac{2}{F_{\alpha\beta\sigma}} \left(\nabla F_{\alpha\beta\sigma}(x,t) \right) \cdot \left(\nabla \delta F_{\alpha\beta\sigma}(x,t) \right) \right\}$$

$$\oplus (-) \left\{ \rho(x,t) \delta F_{\alpha\beta\sigma}(x,t) - \frac{2}{F_{\alpha\beta\sigma}} \left(\nabla F_{\alpha\beta\sigma}(x,t) \right) \cdot \left(\nabla \delta F_{\alpha\beta\sigma}(x,t) \right) \right\} \qquad (1.122)$$

$$\int L_g (ds)^2 \int L_g \left(F_{\alpha\beta\sigma} \right)$$

Given the equations

$$\frac{\delta L_g}{\delta g^{uv}} \to \frac{\delta L_g}{\delta \Gamma^{\alpha\beta}} = \frac{\delta L_g}{\delta \left(\left(\dfrac{1 + 2 F_{uv_{(\alpha)}}}{r^2} \right) \dfrac{dx^2 + dy^2 + dz^2}{dt^2} \right)}$$

$$= \frac{\delta L_g}{\delta \left(\left(\dfrac{1 + 2 F_{uv_{(\alpha)}} M}{r^2} \right) \dfrac{dx^2}{dt^2} + \left(\dfrac{1 + 2 F_{uv_{(\alpha)}} M}{r^2} \right) \dfrac{dy^2}{dt^2} + \left(\dfrac{1 + 2 F_{uv_{(\alpha)}} M}{r^2} \right) \dfrac{dz^2}{dt^2} \right)} \qquad (1.123)$$

$$\frac{\delta L_M}{\delta g^{uv}} \to \frac{\delta L_M}{\delta \Gamma^{\alpha\beta}}$$

$$= \frac{\delta L_M}{\delta \left(\left(\dfrac{1 + 2 F_{uv_{(\alpha)}} M}{r^2} \right) \dfrac{dx^2}{dt^2} + \left(\dfrac{1 + 2 F_{uv_{(\alpha)}} M}{r^2} \right) \dfrac{dy^2}{dt^2} + \left(\dfrac{1 + 2 F_{uv_{(\alpha)}} M}{r^2} \right) \dfrac{dz^2}{dt^2} \right)}$$

The variation of the total Lagrangian takes the form

$$\frac{\delta L}{\delta \Gamma^{\alpha\beta}} = \frac{\delta L_T}{\delta \left(\left(\dfrac{1 + 2 F_{uv_{(\alpha)}}}{r^2} \right) \dfrac{dx^2 + dy^2 + dz^2}{dt^2} \right)}$$

$$= \frac{\delta L_T}{\delta \left(\left(\dfrac{1 + 2 F_{uv_{(\alpha)}} M}{r^2} \right) \dfrac{dx^2}{dt^2} + \left(\dfrac{1 + 2 F_{uv_{(\alpha)}} M}{r^2} \right) \dfrac{dy^2}{dt^2} + \left(\dfrac{1 + 2 F_{uv_{(\alpha)}} M}{r^2} \right) \dfrac{dz^2}{dt^2} \right)} \qquad (1.124)$$

where

$$ds^2 = F_{\alpha\beta}\left(\left(\frac{1+2F_{uv_{(\alpha)}}}{r^2}\right)\frac{dx^2+dy^2+dz^2}{dt^2}\right)$$

$$= \left(\frac{1+2F_{uv_{(\alpha)}}M}{r^2}\right)\frac{dx^2}{dt^2}+\left(\frac{1+2F_{uv_{(\alpha)}}M}{r^2}\right)\frac{dy^2}{dt^2}+\left(\frac{1+2F_{uv_{(\alpha)}}M}{r^2}\right)\frac{dz^2}{dt^2}$$

(1.125)

The length of a curve, in terms of the Lagrangian, and its displacement are given

As

$$L = \int_{\lambda_1}^{\lambda_2}\sqrt{\Gamma_{\alpha\beta}\left(x(\lambda)\left(x^u/d\lambda\right)\left(x^v/d\lambda\right)\right)}d\lambda$$

$$L(\varepsilon) = \int_{\lambda_1}^{\lambda_2}\sqrt{\Gamma_{\alpha\beta}\left(x(\lambda)+\frac{\partial\Gamma_{\beta\beta}}{\partial x^k}\left(x^u/d\lambda\right)\left(x^v/d\lambda\right)\right)}d\lambda$$

(1.126)

Weak gravitational sources

We begin with the equations relating to Newtonian world lines, replaced with the G.R.F.F. potential.

Newtonian world lines read as free-falling particles or inertial particles

$$\left(\frac{\partial}{\partial n}\right)_i = \left(\frac{\partial x^k}{\partial n}\right)\frac{\partial}{\partial x^k} = n^k\frac{\partial}{\partial x^k}$$

(1.127)

Where, in regard to the *G.R.F.F.* potential

$$x^j(t,n)\rightarrow\left(\frac{\partial^2 x_j}{\partial t^2}\right)+\frac{\partial\Phi}{\partial x^j}\rightarrow\left(\frac{\partial^2 x_j}{\partial t^2}\right)+\frac{\partial}{\partial x^j}\left(-F(\gamma)\right)$$

(1.128)

In terms of the Galilean or Newtonian coordinates we will have

$$n+\Delta n\rightarrow\left(\frac{\partial}{\partial n}\right)_i\left[\left(\frac{\partial^2 x_j}{\partial t^2}\right)+\frac{\partial}{\partial x^j}\left(-F(\gamma)\right)\right]$$

$$=\left(\frac{\partial}{\partial n}\right)_i\left(\frac{\partial}{\partial t}\right)_n\left(\frac{\partial x_j}{\partial t}\right)\frac{\partial^2}{\partial x^i\partial x^j}\left(-F(\gamma)\right)$$

(1.129)

Then in a weak gravitational field, nearly inertial coordinate and low velocities, the geodesics take the form of

$$\left(\frac{\partial}{\partial x^{\alpha}}\right) \cdot \left(\frac{\partial}{\partial x^{\beta}}\right) = \eta_{\alpha\beta}$$

$$dx'^{j} / dt'(\gamma) + V = A'_{jk} dx^{k} + a^{j} + V^{j}(t) \tag{1.130}$$

We can write the external field of a weakly gravitating source as

$$\bar{h}_{uv} = h_{uv} = \int \frac{4(\Gamma_{\alpha\beta})(t - |x - x'| x')}{|x - x'|} d^{3}x' = 0 \tag{1.131}$$

Where

$$\Gamma_{\alpha\beta}(t - |x - x'| x') = \sum_{n=1}^{3} \frac{1}{n!} \left[\frac{\partial^{n}}{\partial t^{n}} \Gamma_{\alpha\beta}(t - x, x')\right] = 0 \tag{1.132}$$

And

$$\{dx'^{j} / dt'(\gamma) + V\} \left((dx'^{j} / dt') / r(\gamma) + V\right)$$

$$+ \frac{1}{2} \frac{dx^{j} / dt \cdot dx^{j} / dt}{r} \left(\frac{\{dx^{j} / dt(\gamma) + V\} \cdot ((dx^{k} / dt')(\gamma)) - r'\delta_{jk}}{r^{2}}\right) \tag{1.133}$$

One can determine the conserved total charge of an electromagnetic source by summing the number of electric lines emanating from it, giving us, a Gaussian flux integral over a closed 2-surface, in relation to the *G.R.F.F* potential

$$Q = \frac{1}{4\pi} E^{j} d^{2} S_{j} = \int F d^{2} S_{j} \tag{1.134}$$

If we then define the tidal gravitational forces in correlation to the Riemann curvature tensor, then the relative acceleration of an ensemble or mass of particles, correlates to the tidal producing gravitational force in which we describe and define our central lines of force. The deviation of the particles corresponds to the curvature of the lines of force. We take the conserved momentum and its angular momentum of the stellar body in reference

$$T^{uv}_{,v} = 0 \tag{1.135}$$

Where

$$\int\limits_{body} T^{u0}(t,x)\,dxdydz = P^u = const. \tag{1.136}$$

And

$$\left(\{dx^\alpha / dt(\gamma) + V\} T_{\beta u} - \{dx^\beta / dt(\gamma) + V\} T^{\alpha u} \right)_u \tag{1.137}$$

also

$$\int\limits_{body} \left(\{dx^\alpha / dt(\gamma) + V\} T_{\beta u} - \{dx^\beta / dt(\gamma) + V\} T^{\alpha u} \right) dxdydz = J^{\alpha\beta} = const. \tag{1.138}$$

and

$$P^u = \frac{1}{16\pi} H^{u\alpha 0 j}{}_\alpha dS_j = \frac{1}{16\pi} H^{u\alpha 0 j}{}_j d^3x = \frac{1}{16\pi} H^{u\alpha 0\beta}{}_{\alpha\beta} d^3x = \int T^{u0} d^3x \tag{1.139}$$

We take an isolated system, with a weak gravitational field, which when considering its structure and motion one can disregard self- gravitational effects. We can measure the angular velocity and mass of the gravitational system by measuring the angular velocity and orbital trajectory of test particles within range of its gravitational field. Within the outskirts of the gravitational field the particles are said to move in an elliptical Keplerian orbit. The angular velocity of the source or gravitational system can then be said to written in terms of the precession or angular velocity precession

$$\Omega = \frac{1}{r^3} \left[-S \frac{3(S \cdot x)x}{r^2} \right] \tag{1.140}$$

consider now the mass and angular momentum of gravitating systems such as a star or the sun, and the corresponding gravitational fields, the so called interior or strong region gravitational field and the weak field far from the source. The metric for the weak field can be written as, in correspondence to the gradual decrease in source's mass, angular momentum, and linear momentum, and the non-linearities in the static Newtonian part of the metric, giving us

$$\Gamma_{\alpha\beta} = \left[\left(+\frac{2\left(\Gamma_{uv_{(\alpha)}}\right)M}{r}\right)\delta_{jk}\right]\frac{dx^j dx^k}{dt} \oplus \left[\left(+\frac{2\left(\Gamma_{uv_{(\beta)}}\right)M}{r}\right)\delta_{jk}\right]\frac{dx^j dx^k}{dt}$$

$$\oplus \left[4\varepsilon_{jkl}S^k \frac{x^l}{r^3} + O\left(\frac{1}{r^3}\right)\right]\frac{dx^l}{dt} = +\left[\left(\frac{2\left(\Gamma_{uv_{(\alpha)}}\right)M}{r^2} + \frac{3\left(\Gamma_{uv_{(\alpha)}}\right)M^2}{2r^2}\right)\delta_{jk}\right]\frac{dx^j dx^k}{dt} \qquad (1.141)$$

$$\oplus \left[\left(\frac{2\left(\Gamma_{uv_{(\beta)}}\right)M}{r^2} + \frac{3\left(\Gamma_{uv_{(\beta)}}\right)M^2}{2r^2}\right)\delta_{jk}\right]\frac{dx^j dx^k}{dt} \oplus \left[4\varepsilon_{jkl}S^k \frac{x^l}{r^3} + O\left(\frac{1}{r^3}\right)\right]\frac{dx^l}{dt}$$

We say then that the internal density of the stellar gravitating system, or simply in it's inertial mass, corresponds to the magnitude of its rotating angular momentum, which corresponds proportionally the strength of the interior field, it's length, and entire width of the gravitational field, including the weak field extending from the origin of the source to its limiting extent. The strength of the gravitational fields can be measured by the angular velocity and angle of orbiting velocity of test particles situated within range of the fields. The parameters of the gravitating system, in terms of its mass-density and angular momentum, correspond to the physical dimensions and strengths of the external gravitational fields which corresponds proportionally to the strength of the *G.R.F.F.*, to determined in correlation to the parameters of the orbiting bodies. The flux integrals in the linearized theory can be determined by the linearized field equations. Then in respect to the previously defined notation our equations for conservation of total 4-momentum and angular momentum read

$$P^u = \sum_A \int_A T^{u0}d^3x + \int_{\substack{inter \\ body \\ region}} T^{u0}d^3x = \sum_A \int_A P^u_A d^3x + \int_{\substack{inter \\ body \\ region}} T^{u0}d^3x \qquad (1.142)$$

Where

$$J^{uv} = \sum_A J^{uv}_A + \int_{\substack{inter \\ body \\ region}} \left(x^u T^{v0} - x^v T^{u0}\right)d^3x \qquad (1.143)$$

With

$$\frac{dP^{uv}}{dt} = \frac{d}{dt}\int T^{u0}d^3x = \int T_0^{u0}d^3x = \int_A T_f^{uf}d^3x \int_A T^{uf}d^2S_j \qquad (1.144)$$

And

$$\frac{d}{dt}J^{uv} = \int_{S_2}\left(x^u T^{vj} - x^v T^{uj}\right)d^2S_j \qquad (1.145)$$

Hilbert variational principle

The Hilbert variational principle can be written as

$$I = \int Ld^3x = \int L\left(\Gamma_{\alpha\beta}\right)d^3x \qquad (1.146)$$

Giving us then

$$I = \int Ld^3x = \int L\left(T_{uv}\right)^{1/2}d^3x = \int_{dx,dt}^{x,t} L\left(dx/dt, x, t\right) = L_M \qquad (1.147)$$

the variational principle can be written as

$$dI = \int_{dx,dt}^{x,t}\left[\left\{p(t)\frac{dx(t)}{dt} \pm \frac{d}{dt}\left(p(t)\frac{dx(t)}{dt}\right)\right\} - dH\left((t), x(t), t\right)\right]dt \qquad (1.148)$$

or written in respect to moving frames of reference our variation becomes

$$dI = \int_{dx,dt}^{x,t}\left[\left\{p(t)\frac{dx(t)}{dt} \pm \frac{d}{dt}(T)\left(p(t)\frac{dx(t)}{dt}\right)\right\} - dH\left((t), x(t), t\right)\right]dt \qquad (1.149)$$

when studying the structure and dynamics of massive stellar relativistic stars, we first concern ourselves with the metric before deducing the equations for the gravitational field of a static spherical star, giving us

$$A = \int\left(rd\theta\left(r\sin\theta d\phi\right)\right) = 4\pi r^2 \qquad (1.150)$$

Internal structure of stars

If we were to analyze the internal structure of a stellar object we would find the matter within the body to be describable in relation to a perfect fluid. Then considering this we use the physical parameters in analogy to a perfect fluid. If the star is static then each element of material composite of the star is at rest in its given frame of reference, or static coordinate system. Each unit element must move according to the equations relating its velocity in polar coordinates, we can then obtain the formula for the stress-energy tensor. The internal structure is then said to be determined by the field equations and the local conservation of energy-momentum present. Included are the functionals describing the value of internal pressure and density, in correlation to the number density of baryons. Then we can determine the entropy of energy per baryon. Some physical examples of relativistic stars include, neutron stars, white dwarfs, and supermassive stars, each existing in theory. A star is said to collapse from its initial state to a neutron star under required conditions. The structure of the star or massive stellar body can then be said to consist of the five functions of the radius, giving us the metric function, the pressure, density of mass-energy, and density of baryons. If we consider these initial conditions along with the total mass-energy inside the star or within a given radius, for a spherically symmetric star, then centered at the core where it is found to be densest, it decreases in density proportionally in a geometric progression, as we move away from the core. Then we likewise assume the body to be generally uniform in structure and uniform in its distribution of matter. The density varies depending on the range or location within the star. The diameter of the star corresponds to the size and density of its core.

The larger the diameter, generally, the larger and denser the core, and correspondingly the larger and stronger the gravitational field, or external field. If our stellar body were fixed in space relative to neighboring stars then we can concern ourselves with its internal dynamics, that is its intrinsic rotation, and angular momentum. Where the structure of the star, that is its internal structure is determined by the five functions of the radius r. We concern ourselves with the value of the pressure gradient, required to prevent any fluid element from leaving the internal radius of the star. We then derive a balance between the gravitational force and the pressure gradient. As we near the center of the star we find the pressure, internal pressure, increases proportional to diameter of the star and its coreIf our body were static and spherically symmetric, and for the moment non-rotating, such that the values for its angular velocity vanish, then we find the magnitude and structure of its corresponding external gravitational field corresponds to the structure of the field.

Structure corresponding to the G.R.F.F. potential

The lines of force extending from the center of origin of the stellar body, as determined by the number of particles and physical energy leaving and the rate at which they leave, before generally dissipating into space, corresponds to their physical strength as determined by the number of lines crossing per unit volume, as depicted in polar coordinates within range of the field, and the length of the lines extending away from the body. If the flow of energy, leaving the body, was constant, then we could determine the range of the gravitational field and its physical strength. We generally assume that once we get closer to the star the strength of the gravitational field, as determined by the magnitude of the linear momentum and density of energy, fields, and particles leaving the body, grows in strength. The strength then is determined as by the physical parameters previously introduced. If the lines of force as determined by the trajectory or geodesics of the particles leaving the star are equivalent to that of the charged particle, then we say the gravitational field is divergent, with each line extending equally in each direction. The means in which material within the strong region of the gravitational field, orbits around the body, considering some to dissipate due to the heat or and fall into the star, as it gathers energy in the form of particles and debris, corresponds to the means in which all material that either falls into the star, (not being pulled in due to the internal gravitational field but due to the body's inability to remain a constant orbiting trajectory) or continuous along an elliptical motion or trajectory. All bodies that initially move along straight lines or geodesics are said to move now in elliptical paths along curved trajectories or geodesics due in fact to the conditions previously defined for all orbiting bodies. Then the bodies in question are said to either be moving out of physical range of the star's gravitational field, falling into the star, or moving about the star along an elliptical Keplerian orbital, whose parameters are defined by the physical conditions previously described. Then in describing orbiting bodies in respect to the source or center of gravity, we concern ourselves neither with the structure of space nor the extending or curved lines of force, but rather the condition in which a body would tend to move along a curve or elliptical path rather than its initial linear path of motion, due in fact to the conditions previously defined.

If we write out the form of the curvature tensor and transformation properties, we then derive the variation of the curvature tensor and Lagrangian density

$$\Gamma^{\lambda}_{\alpha\beta} = \left[\Gamma^{\lambda}_{\sigma\tau} \frac{\partial x^{\sigma}}{\partial x^{\alpha}} \frac{\partial x^{\tau}}{\partial x^{\beta}} + \frac{\partial^2 x^{\lambda}}{\partial x^{\alpha} \partial x^{\beta}} \right] \frac{\partial x^{\gamma}}{\partial x^{\lambda}} \tag{1.151}$$

Action principle

For the extremum of the action principle to exist the following relation must hold valid

$$\int \left[\Gamma_{\alpha\beta} \left(d\Gamma^{\lambda}_{uv,\lambda} - d\Gamma^{\lambda}_{u\sigma,\lambda} \right) \right] \Gamma_{\alpha\beta} d^3 x + \int \left(\frac{\partial L_M}{\partial \Gamma_{\alpha\beta}} - \frac{1}{2} \Gamma_{\alpha\beta} L_M \right) d\Gamma_{\alpha\beta} d^3 x \tag{1.152}$$

Where

$$T_{\alpha\beta} = 2 \frac{\partial L_M}{\partial \Gamma_{\alpha\beta}} - \frac{1}{2} \Gamma_{\alpha\beta} L_M \tag{1.153}$$

And

$$\Gamma_{\alpha\beta} T^{\alpha\beta} = \mathfrak{I}^{\alpha\beta} = 2 \frac{\partial L_M}{\partial \Gamma_{\alpha\beta}} \tag{1.154}$$

For intrinsic curvature to the 3-geometry, we consider the displacement, and its scalar product

$$dP = e_j dx^j = A = e_j A^l = \left(A \cdot e_j \right) = A^i \left(e_i \cdot e_j \right) = A^i g_{ij} = A_j \qquad (1.155)$$

is then given as

$$\nabla_{e_i} A = \nabla_i A = \nabla \left(e_j A^l \right) = e_j \frac{\partial A^j}{\partial x^i} + \left(\Gamma^u_{ji} e_u \right) A^j \qquad (1.156)$$

and the local law of conservation of energy is given as

$$\frac{d\rho}{dt} = -\left(\rho + p \right) \nabla \cdot u = \frac{\rho + p}{nt} \frac{dn}{dt} = -\left(\rho + p \right) \nabla_u = -\left[\nabla p + \left(\nabla_u p \right) u \right]$$
$$= -\left(\rho + p \right) u_{r;v} u^v = -\left(\rho + p \right) \Gamma^\alpha_{rv} u_\alpha u^v = -\left(\rho + p \right) \Gamma^0_{r0} u_0 u^0 = -p_r \qquad (1.157).$$

External gravitational field

With the external gravitational field, the density and pressure outside the star vanish and the space and time geometry equal zero, while the metric in terms of the motion of freely falling bodies becomes

$$\Gamma_{\alpha\beta} = \frac{dr^2}{\left(1 - \left\{ \frac{2\left(F_{uv_{(\alpha)}} \right) Mm}{r^2} \right\} \right)} \oplus \frac{dr^2}{\left(1 - \left\{ \frac{2\left(F_{uv_{(\beta)}} \right) Mm}{r^2} \right\} \right)} + r^2 \left(d\theta + \sin^2 \theta d\phi^2 \right) \qquad (1.158)$$

The endpoint of stellar evolution, or were the star is said to die, occurs in a variety of ways. One, either the star dissipates into space or the interstellar medium, contracts into a white-dwarf state or neutron-star state, or into a black hole. In determining the final state of stellar evolution, a star, an active star, consisting of any given number of baryons, reaches an absolute, burned-out point of thermonuclear combustion, derived as the state of minimum mass-energy possible to the given baryon system. Once the system arrives at a state of burnout, by collecting outgoing matter, extracting its kinetic energy and falling back into the system, the system will reach its lowest energy state, with zero angular momentum, and temperature, having all physical heat removed, ending in a cold stellar configuration. In this state, the equations of state can be characterized by zero thermonuclear activity, zero dissipation and creation of particles or baryons, zero mass and energy extraction and dissipation, and limited particles or material activity within the core. Once the cold core or center has been compressed to high enough densities, the parameters previously defined in determining the functional state of the star vanish, resulting in the cold stellar configuration. Corresponding to every central density value, we have a single stellar equilibrium configuration.

The motion of any material body or particle moving in the external gravitational field of a static star, is defined by the line element redefined as

$$ds^2 = \frac{dr^2}{\left(1-2\left(\Gamma_{uv_{(\alpha)}}\right)\right)} \oplus \frac{dr^2}{\left(1-2\left(\Gamma_{uv_{(\beta)}}\right)\right)} + r^2\left(d\theta + \sin^2\theta d\phi^2\right) \tag{1.159}$$

Whereas was shown the vector of the metric is curved and not the physical space or time. The path of motion remains the same while the physical motion itself is curved in correspondence to the elliptical motion of the body. This applies to all stellar stars of any spherically symmetric center of attraction, with a denoted total mass-energy M. The given test body or particle, regardless of angular velocity, can be defined by a fixed point and 3-velocity projected radially onto a sphere of a fixed value denoted r. They define the total circular motion of the test body or particle. The radial projection of its relative position will continue to lie on the defined circle. We then orient the chosen coordinate system so that the radial projection of the orbit coincides with the equator. In polar coordinates the particle has zero momentum in the given direction

$$p^\theta = d\theta/d\lambda = 0 \tag{1.160}$$

the 3-vector of the particle's energy-momentum is given as

$$\Gamma_{\alpha\beta} p^{\alpha} p^{\beta} + u^2 = \Gamma_{\alpha\beta} P_{\alpha} P_{\beta} + u^2 = \left\{ \left(\frac{E^2}{\left(1-2\left(\Gamma_{uv_{(\alpha)}}\right)\right)\frac{Mm}{r^2}} \right) \oplus \left(\frac{E^2}{\left(1-2\left(\Gamma_{uv_{(\beta)}}\right)\right)\frac{Mm}{r^2}} \right) \right\}$$

$$+ \left\{ \left(\frac{1}{\left(1-2\left(\Gamma_{uv_{(\alpha)}}\right)\right)\frac{Mm}{r^2}} \right) \oplus \left(\frac{1}{\left(1-2\left(\Gamma_{uv_{(\beta)}}\right)\right)\frac{Mm}{r^2}} \right) \right\} \left(\frac{dr}{d\lambda} \right) + \frac{L^2}{r^2} + u^2 \tag{1.161}$$

Where

$$\left(\frac{dr}{d\lambda} \right)^2 = E^2 - \left\{ \left(\frac{1}{\left(1-2\left(\Gamma_{uv_{(\alpha)}}\right)\right)\frac{Mm}{r^2}} \right) \oplus \left(\frac{1}{\left(1-2\left(\Gamma_{uv_{(\beta)}}\right)\right)\frac{Mm}{r^2}} \right) \right\} \left(1 + L^2 / r^2\right) \tag{1.162}$$

$$= E^2 - V^2(r)$$

which defines the orbital motion of the test particle or body. The last equation relates the effective potential. To define the proper time for the particle to complete its orbital or orbit in our new post-Newtonian coordinates we have, the latter equation relating the falling particle or particle with zero velocity and the particle's given motion,

$$\tau = \int d\tau = \frac{dr}{\sqrt{E^2 \left\{ \left(\frac{1}{\left(1-2\left(\Gamma_{uv_{(\alpha)}}\right)\right)\frac{Mm}{r^2}} \right) \oplus \left(\frac{1}{\left(1-2\left(\Gamma_{uv_{(\beta)}}\right)\right)\frac{Mm}{r^2}} \right) \right\} \left(1 + L^2 / r^2\right)}}$$

$$= \int \frac{dr}{\sqrt{\left(2\left(\Gamma_{uv_{(\alpha)}}\right)\frac{Mm}{r^2} \right) \oplus \left(2\left(\Gamma_{uv_{(\beta)}}\right)\frac{Mm}{r^2} \right)}} \tag{1.163}$$

And

$$\frac{dr}{d\tau} = \frac{dr}{dt}\frac{dt}{d\tau} = \frac{dr}{dt}\left(\frac{E}{\left(\dfrac{1}{\left(1-2\left(\Gamma_{uv_{(\alpha)}}\right)\right)\dfrac{Mm}{r^2}}\right) \oplus \left(\dfrac{1}{\left(1-2\left(\Gamma_{uv_{(\beta)}}\right)\right)\dfrac{Mm}{r^2}}\right)}\right) = E\frac{dr^*}{dt} \qquad (1.164)$$

the time coordinate of the moving body or test particle can be given as

$$t = \int dt = \int \frac{E\,dr}{\sqrt{E^2 - V^2}}$$

$$= \int \frac{dr}{\sqrt{E^2 - \left\{\left(2\left(\Gamma_{uv_{(\alpha)}}\right)\dfrac{Mm}{r^2}\right) \oplus \left(2\left(\Gamma_{uv_{(\beta)}}\right)\dfrac{Mm}{r^2}\right)\right\}\left(1+L^2/r^2\right)}} \qquad (1.165)$$

$$\frac{dr}{\left(\dfrac{1}{\left(1-2\left(\Gamma_{uv_{(\alpha)}}\right)\right)\dfrac{Mm}{r^2}}\right) \oplus \left(\dfrac{1}{\left(1-2\left(\Gamma_{uv_{(\beta)}}\right)\right)\dfrac{Mm}{r^2}}\right)}$$

And

$$\frac{dr}{d\tau} = \frac{dr}{dt}\frac{dt}{d\tau} = \frac{dr}{dt}\left(\frac{E}{\left(\dfrac{1}{\left(1-2\left(\Gamma_{uv_{(\alpha)}}\right)\right)\dfrac{Mm}{r^2}}\right) \oplus \left(\dfrac{1}{\left(1-2\left(\Gamma_{uv_{(\beta)}}\right)\right)\dfrac{Mm}{r^2}}\right)}\right) = E\frac{dr^*}{dt}$$

$$r^* = \int dr^* = \int \frac{dr}{\left(\dfrac{1}{\left(1-2\left(\Gamma_{uv_{(\alpha)}}\right)\right)\dfrac{Mm}{r^2}}\right) \oplus \left(\dfrac{1}{\left(1-2\left(\Gamma_{uv_{(\beta)}}\right)\right)\dfrac{Mm}{r^2}}\right)} \qquad (1.166)$$

$$= r + 2M\ln\left(\left(2\left(\left(\Gamma_{uv_{(\alpha)}}\right)\dfrac{Mm}{r^2}\right)-1\right) \oplus \left(2\left(\Gamma_{uv_{(\beta)}}\right)\dfrac{Mm}{r^2}-1\right)\right)$$

Lagrangian and Hamiltonian Forms

The Lagrangian and Hamiltonian forms of the equations of motions may be written as

$$\delta \int_{t_1}^{t_2} L\left(u^i, x^i, t\right) dt = 0 \qquad (1.167)$$

Where the Lagrangian is given as

$$L = m_0 \sqrt{\left(1 + 2 \frac{\left(F_{uv_{(\alpha)}} \oplus F_{uv_{(\beta)}}\right) M}{r^2} \frac{dx^2 + dy^2 + dz^2}{dt^2}\right)} \qquad (1.168)$$

Here we will have the gravitational potential in terms of the *G.R.F.F.* potential, as the line element and spatial line element,

$$(dv)^2 = \frac{\left(F_{uv_{(\alpha)}}\right) M}{r^2} \frac{dx^i}{dt} \oplus \frac{\left(F_{uv_{(\beta)}}\right) M}{r^2} \frac{dx^j}{dt} \qquad (1.169)$$

then as

$$(dv)^2 = \left(1 + \frac{\left(F_{uv_{(\alpha)}} \oplus F_{uv_{(\beta)}}\right) M}{r^2}\right) \frac{dx^2 + dy^2 + dz^2}{dt^2} \qquad (1.170)$$

Take an arbitrary system of coordinates, with the given metric tensor, giving us the transformed components of the metric tensor in terms of the new tensor. We characterize the space and time components as form variant, as in the case of permanent gravitational fields, under a group of variant orthogonal transformations. These variants, of the group of three-dimensional rotations in Euclidean space and the line element in a spherical system, is given as

$$\frac{dx^2 + dy^2 + dz^2}{dt^2} = \frac{dr^2 + r^2 d\theta + r^2 \sin^2 \theta d\phi^2}{dt^2} \qquad (1.171)$$

giving us then

$$dv = F_{uv} \frac{(r,t)dr^2}{dt^2} + F_{uv}(r,t)\frac{dr^2 + r^2 d\theta + r^2 \sin^2 \theta d\phi^2}{dt^2}$$
$$= F_{uv} \frac{(r',t)dr'^2}{dt^2} + F_{uv}(r',t)\frac{dr'^2 + r'^2 d\theta + r'^2 \sin^2 \theta d\phi^2}{dt^2}$$

(1.172)

now in determining the dynamical gravitational potential, or the strength of the gravitational field, which neglecting the structure of the field itself is concerned primarily with what originally causes the deviation. We consider the use of a test body or particles, situated at rest within a point of our chosen coordinate system where the gravitational field is to be measured. We assume the magnitude of the particle's acceleration due not to the lines of force of the gravitational field which would originally determine the strength of the fields lines of force and field itself.

We take our reformulation of Newton's law of gravity between two bodies of varying mass such that the smaller mass is more affected by the presence of the larger mass. Then according to the law of action and reaction, the larger mass is largely undisturbed by the presence of the significantly smaller mass.

9

Surface-gravity as a frame-dependent phenomenon: modern formulation

Introduction

We employ modern mathematical formulism in defining the surface-gravitational field as a rotating frame-dependent phenomenon.

If we define the Earth's gravitational field in terms of its intrinsic rotation, the linear deviation of the lines of force (virtual; not real) that extend a finite distance uniformly around the circumference of the body, corresponds to the inertial mass of the rotating body, and the angular velocity at which it rotates. The rotation itself acts uniformly on all material within its range, that is within the range of the force induced by its rotation. Since in general relativity mass is the key source to gravity we can represent the body's mass by a stress-energy tensor.

If we define the direction of the lines of force (fictitious: virtual lines) of the force field under consideration, then the set vectors positioned perpendicular to the surface of the earth can define the forces acting on all bodies due to the Earth's rotation. The length and number of representational lines crossing any unit area positioned relative to and parallel to the surface of the body of reference determines the net strength of that unit area uniformly on all physical material with that directed force, all directly proportional to the magnitude of the Earth's rotation. If we disregard external retarding factors such as wind, we can consider solely the nature of these lines of force, representing the force due to the Earth's rotation, in correlation to the motion of bodies under its influence. That is to say, if we have a spherical mass or oblate mass of equal mass distribution of say significant density, then its position in space can be determined relative to the stars in a polar coordinate system. If we construct a series of polar coordinate frames on the lattice or physical structure, then we will find the variance of the directed, virtual, vector fields moving parallel to the surface of the lattice and horizontal through the defined polar coordinates. If the body of reference is at rest within the designated Cartesian frame, and is not rotating with any given angular momentum, then the lines of force would not be useful. Than any body of significant mass positioned parallel and on the surface of the body, unless acted upon otherwise by any frictional or static forces, that hold the body in a fixed positioned relative to the surface, then the state of the body or bodies is determined by the state of the medium in which in it is depicted relative to the surface of the spherical lattice. In this case, the virtual vector fields consists of the lines of force, in which their direction and magnitude would be determined by rotational momentum of the lattice structure, and their polar divergence would keep any body from staying in a fixed position relative to the surface of the lattice. Then anybody projected from the surface would move in a straight line in the direction of the lines of force. The body would tend to move in the same direction as the lines of force. There would be no relative curvature of its trajectory relative to the surface of the lattice.

Now if we consider the analogy of the rotating charge, rotating on its axis with a constant angular velocity fixed within Cartesian coordinates, then we would measure no divergence of the lines of force emanating from the charge. The lines of force would thusly take on a curved trajectory and move in a direction parallel to the rotation of the charge, taking a helix formation. The effect of the rotation of the charge would affect the direction of the lines of force emanating from it. If we consider this analogy and apply it to all rotating bodies, that is all orbiting rotating bodies, we find a possible shift in the original linear trajectory of bodies depicted with the polar frames on the surface of the rotating lattice. Here the lines are purely fictitious. If the rotation is constant and the shape approximately spherical, then the force of its rotation, in terms of non-centrifugal forces would act uniformly on all bodies situated within its physical range, that is the range of the field of force, fixing them in a defined positioned relative to and on the rotating lattice. It would also explain the curvature and parabolic trajectories and free fall of these same bodies situated within the field of force, the so named gravitational field of force. All motion in terms of the classical laws are then said to operate in respect to this field of force. The origin and determination of the lines of force are then said to be a result of the rotation of the body and not solely from its innate mass.

If we have a body set in uniform motion that is resisted by a resisting force in ratio to its initial velocity, the motion lost due to the resistance is proportional to the space traversed by the moving body. The body in motion, the amount in which its original motion is retarded in correlation to its terminal velocity, is proportional to the magnitude of the resisting force or the initial characteristics of the medium in which the body moves through. As was stated the resistance follows a geometrical progression as does the change in velocity and times. Anybody then descending or ascending in a straight line in a uniform medium where external resistance, i.e. air, is neglected, then the resistance due to the uniform force of gravity can be defined in relation and relative to the surface and motion of the frame of reference. The massive spherical body in which we define polar coordinates on and relative to its surface, its intrinsic rotation and orbiting velocity multiplied by its mass, correspond to the magnitude of the lines of force of which gravity manifests itself, within the chosen coordinate frames. The direction and magnitude of the force acts uniformly on any and all bodies situated within the frame of reference corresponds to the position and structure of the massive body's surface. The times in which a body descends or ascends in correlation to any external applied force that would accelerate the body, that is to move it in a straight or parabolic line relative to the surface of the planetary body corresponds to the mass of the body in motion, the applied force, and its magnitude. If the total collection of forces (surface gravity forces), aside from the action of the medium, are to move in a geometrical progression, then a projectile set in relative motion, relative to the surface as projected from it by a finite force, will move under this progression as it ascends by the force of the projected velocity and the resistance of the medium, following a curve throughout its ascension, moving approximately perpendicular (with an angle as determined by the angle it was projected from and the force it was projected by) to the direction of the body's surface. They move parallel to the surface of the earth, and will reach relative maximum height before beginning descension. The key height it will reach is determined conjunctly both by the magnitude of the body's rotational velocity and the angle of projection and force of projection. The descension is defined then conjunctly by these parameters, descending at an angle and rate proportional to the angle of and force of ascension and the force through which it moves. The curve described by the bodies motion can be defined and described by hyperbolas and parabolas whose angles and sectors depend on the physical parameters previously described.

The surface gravity, the rotationally-dependent field of force, given by the intrinsic rotation of the smaller mass will be fuller defined by the following equations

$$mf = -F(\beta) + m\omega \times (\omega \times r) + 2m\omega \times v = -F_g(\beta) + m\omega \times (\omega \times r) + 2m\omega \times v$$

$$\Rightarrow \sum_{i=1}^{n} mf = \sum_{i=1}^{n} \left(-F(\beta) + m\omega \times (\omega \times r) + 2m\omega \times v \right) \tag{1.173}$$

$$= \sum_{i=1}^{n} \left(-F_g(\beta) + m\omega \times (\omega \times r) + 2m\omega \times v \right)$$

where

$$-F_g(\beta) + m\omega \times (\omega \times r)$$
$$f = -g + 2\omega \times v \tag{1.174}$$

and

$$\sum_{i=1}^{n} (\omega \times r) = \sum_{i=1}^{n} i\left(w_y z - w_z y\right) + \sum_{i=1}^{n} i\left(w_y z - w_x y\right) + \sum_{i=1}^{n} i\left(w_y z - w_y y\right) \tag{1.175}$$

where the sum is the motion of all collected particles and related materials moving relative to the surface of the larger rotating body, constituting that special vector field as determined by the rotationally-dependent frame of reference.

We can now integrate over a unit area within the defined vector field as defined by the deviation of the particles moving according to the above equation. We can write then

$$\int_\tau (V + (\omega \times r)) d\tau = i \int_\tau (V_x + (\omega \times r)) d\tau + j \int_\tau (_y V + (\omega \times r)) d\tau$$
$$+ k \int_\tau (V_z + (\omega \times r)) d\tau$$

$$\int_\tau (V_{x_1} + (\omega \times r)) d\tau = \int_\tau (V_{x_1} + (\omega \times r)) d\tau + \int_\tau (_{y_1} V + (\omega \times r)) d\tau$$
$$+ \int_\tau (V_{z_1} + (\omega \times r)) d\tau + ...,$$

$$\int_\tau (V_{x_2} + (\omega \times r)) d\tau = \int_\tau (V_{x_2} + (\omega \times r)) d\tau + \int_\tau (_{y_2} V + (\omega \times r)) d\tau$$
$$+ \int_\tau (V_{z_2} + (\omega \times r)) d\tau + ...,$$

$$\int_\tau (V_{x_3} + (\omega \times r)) d\tau = \int_\tau (V_{x_3} + (\omega \times r)) d\tau + \int_\tau (_{y_3} V + (\omega \times r)) d\tau$$
$$+ \int_\tau (V_{z_3} + (\omega \times r)) d\tau + ...,$$

(1.176)

Which can be written in terms of Gauss's theorem for unit volumes, giving us

$$\sum (V + (\omega \times r)) \cdot d\sigma = \int_\sigma (V + (\omega \times r)) \cdot d\sigma$$
$$= \int_\sigma (V_x + (\omega \times r)) \cdot d\sigma + \int_\sigma (V_y + (\omega \times r)) \cdot d\sigma + \int_\sigma (V_z + (\omega \times r)) \cdot d\sigma$$

(1.177)

Then

$$\sum (\nabla \cdot V_x + (\omega \times r)) d\tau + \iiint \left\{ \frac{\partial V_x + (\omega \times r)}{\partial x} + \frac{\partial V_y + (\omega \times r)}{\partial y} + \frac{\partial V_z + (\omega \times r)}{\partial z} \right\}$$
$$= \iiint \frac{\partial V_x + (\omega \times r)}{\partial x} dxdydz + \iiint \frac{\partial V_y + (\omega \times r)}{\partial y} dxdydz + \iiint \frac{\partial V_z + (\omega \times r)}{\partial z} dxdydz$$

(1.178)

Followed by

$$\iiint \frac{\partial V_x + (\omega \times r)}{\partial x} dxdydz = \iint \left\{ \left(V_x + (\omega \times r) \right)(x_2) - \left(V_x + (\omega \times r) \right)(x_2) \right\} dydz$$

$$= \int_{\sigma} \left(V_x + (\omega \times r) \right) d\sigma_x$$

$$\iiint \frac{\partial V_y + (\omega \times r)}{\partial x} dxdydz = \iint \left\{ \left(V_y + (\omega \times r) \right)(y_2) - \left(V_y + (\omega \times r) \right)(y_2) \right\} dxdz$$

$$= \int_{\sigma} \left(V_y + (\omega \times r) \right) d\sigma_y$$

$$\iiint \frac{\partial V_z + (\omega \times r)}{\partial x} dxdydz = \iint \left\{ \left(V_y + (\omega \times r) \right)(z_2) - \left(V_z + (\omega \times r) \right)(z_2) \right\} dxdy$$

$$= \int_{\sigma} \left(V_z + (\omega \times r) \right) d\sigma_z$$

(1.179)

And finally

$$\int_{\sigma} \left[\nabla \cdot \left(V_x + (\omega \times r) \right) d\tau \right]$$

$$= \int_{\sigma} \left(\left(V_x + (\omega \times r) \right) d\sigma_x + \left(V_y + (\omega \times r) \right) d\sigma_y + \left(V_z + (\omega \times r) \right) d\sigma_z \right)$$

(1.180)

72

10

Hamilton-Jacobi method

Introduction

Here we focus on the G.R.F.F potential, in temrs of the Hamilton-Jacobi method.

For the motion under the influence of the *G.R.F.F.* field analyzed by the Hamilton-Jacobi method we will have

$$H = \frac{p_r^2}{2} + \frac{p_r^2}{2r\left(\Gamma_{uv_{(\alpha)}} \oplus \Gamma_{uv_{(\beta)}}\right)} + \frac{p_r^2}{2r\left(\Gamma_{uv_{(\alpha)}} \oplus \Gamma_{uv_{(\beta)}}\right)\sin^2\theta} - r\left(\Gamma_{uv_{(\alpha)}} \oplus \Gamma_{uv_{(\beta)}}\right) \tag{1.181}$$

and

$$S = -Et + p_\phi\phi + \int^\theta\left(L^2 - \frac{p_\phi^2}{\sin^2\theta}\right)d\theta + \int^r\left[2\left(E + r\left(\Gamma_{uv_{(u)}} \oplus \Gamma_{uv_{(\beta)}}\right) - \frac{L^2}{2r^2}\right)\right]dr + \delta_{p_\phi L,E}$$

$$= -Et + L\theta + \int^r\left[2\left(E + r\left(\Gamma_{uv_{(\alpha)}} \oplus \Gamma_{uv_{(\beta)}}\right) - \frac{L^2}{2r\left(\Gamma_{uv_{(\alpha)}} \oplus \Gamma_{uv_{(\beta)}}\right)^2}\right)\right]dr + \delta_{p_\phi L,E} \tag{1.182}$$

followed by

$$-\frac{\partial S}{\partial t} = \frac{1}{2}\left(\frac{\partial S}{\partial r}\right)^2 + \frac{1}{2r^2}\left(\frac{\partial S}{\partial r}\right)^2 + \frac{1}{2r^2\sin^2\theta}\left(\frac{\partial S}{\partial r}\right) + r\left(\Gamma_{uv_{(\alpha)}} \oplus \Gamma_{uv_{(\beta)}}\right) \tag{1.183}$$

then

$$\frac{\partial S}{\partial t} = \theta + \int^r \frac{Ldr / r^2}{2\left(E + r\left(\Gamma_{uv_{(\alpha)}} \oplus \Gamma_{uv_{(\beta)}}\right) - \frac{L^2}{2r\left(\Gamma_{uv_{(\alpha)}} \oplus \Gamma_{uv_{(\beta)}}\right)^2}\right)} \qquad (1.184)$$

followed by a similar relation

$$\frac{\partial S}{\partial E} = t + \int^r \frac{Ldr / r^2}{2\left(E + r\left(\Gamma_{uv_{(\alpha)}} \oplus \Gamma_{uv_{(\beta)}}\right) - \frac{L^2}{2r\left(\Gamma_{uv_{(\alpha)}} \oplus \Gamma_{uv_{(\beta)}}\right)^2}\right)} \qquad (1.185)$$

the magnitude, in terms of the *G.R.F.F.* potential, of the momentum of the particle moving in this potential is given as

$$\Gamma_{\alpha\beta} p^{\alpha} p^{\beta} + u^2 = \Gamma_{\alpha\beta} p_{\alpha} p_{\beta} + u^2$$

$$= \frac{E^2}{\left(1 + 2r\left(\Gamma_{uv_{(\alpha)}} \oplus \Gamma_{uv_{(\beta)}}\right)\right)} + \frac{1}{\left(1 + 2r\left(\Gamma_{uv_{(\alpha)}} \oplus \Gamma_{uv_{(\beta)}}\right)\right)} \left(\frac{dr}{d\lambda}\right)^2$$

$$+ Lr^2 \left(\Gamma_{uv_{(\alpha)}} \oplus \Gamma_{uv_{(\beta)}}\right)^2 + u^2 \qquad (1.186)$$

$$= \left(\frac{dr}{d\lambda}\right)^2 = E^2 + \left(1 + 2r\left(\Gamma_{uv_{(\alpha)}} \oplus \Gamma_{uv_{(\beta)}}\right)\right)\left(1 + Lr^2\left(1 + 2r\left(\Gamma_{uv_{(\alpha)}} \oplus \Gamma_{uv_{(\beta)}}\right)^2\right)\right)$$

$$= E^2 + \sqrt{\left(1 + 2r\left(\Gamma_{uv_{(\alpha)}} \oplus \Gamma_{uv_{(\beta)}}\right)\right)\left(1 + Lr^2\left(1 + 2r\left(\Gamma_{uv_{(\alpha)}} \oplus \Gamma_{uv_{(\beta)}}\right)^2\right)\right)}$$

in terms of the angle ϕ we will have

$$\frac{d\phi}{d\tau} = \frac{1}{u}\frac{d\phi}{d\lambda} = \frac{p^{\phi}}{u} = \frac{L}{r^2} \qquad (1.187)$$

and then in terms of the variable t

$$\frac{dt}{d\tau} = \frac{1}{u}\frac{dt}{d\lambda} = \frac{p^0}{u} = \frac{E}{1 - 2r\left(\Gamma_{uv_{(\alpha)}} \oplus \Gamma_{uv_{(\beta)}}\right)} \qquad (1.188)$$

74

Schwarzschild Geometry rewritten in terms of the G.R.F.F. potential

We begin with the time component

$$\frac{dr}{d\tau} = \frac{dr}{dt}\frac{dt}{d\lambda} = \frac{dr}{dt} = \frac{E}{1+2r\left(\Gamma_{uv_{(\alpha)}}\oplus\Gamma_{uv_{(\beta)}}\right)} = E\frac{dr}{dt} \tag{1.189}$$

taking the square gives us

$$\left(\frac{dr}{d\tau}\right)+\left(1+2r\left(\Gamma_{uv_{(\alpha)}}\oplus\Gamma_{uv_{(\beta)}}\right)\right)\left(1+Lr^2\left(1+2r\left(\Gamma_{uv_{(\alpha)}}\oplus\Gamma_{uv_{(\beta)}}\right)^2\right)\right) \tag{1.190}$$

we use the lower time part and write it as an integral, giving us

$$\tau = \int d\tau = \int \frac{dr}{\left[E^2+\left(1+2r\left(\Gamma_{uv_{(\alpha)}}\oplus\Gamma_{uv_{(\beta)}}\right)\right)\left(1+Lr^2\left(1+2r\left(\Gamma_{uv_{(\alpha)}}\oplus\Gamma_{uv_{(\beta)}}\right)^2\right)\right)\right]}$$

$$= \int \frac{dr}{\left[2r\left(\Gamma_{uv_{(\alpha)}}\oplus\Gamma_{uv_{(\beta)}}\right)\right]^{1/2}} = \frac{R}{2}\left(\frac{R}{2M}\right)^{1/2}(\eta+\sin\eta)\frac{Edr}{\left[E^2-V^2\right]^{1/2}}$$

$$= \int \frac{dr}{\left[E^2+\left(1+2r\left(\Gamma_{uv_{(\alpha)}}\oplus\Gamma_{uv_{(\beta)}}\right)\right)\left(1+Lr^2\left(1+2r\left(\Gamma_{uv_{(\alpha)}}\oplus\Gamma_{uv_{(\beta)}}\right)^2\right)\right)\right]^{1/2}}$$

$$\left(\frac{dr}{\left(1+2r\left(\Gamma_{uv_{(\alpha)}}\oplus\Gamma_{uv_{(\beta)}}\right)\right)}\right)$$

$$= \int \frac{\left[1+2R\left(\Gamma_{uv_{(\alpha)}}\oplus\Gamma_{uv_{(\beta)}}\right)\right]^{1/2}}{\left[2R\left(\Gamma_{uv_{(\alpha)}}\oplus\Gamma_{uv_{(\beta)}}\right)+2r\left(\Gamma_{uv_{(\alpha)}}\oplus\Gamma_{uv_{(\beta)}}\right)\right]^{1/2}}+\frac{dr}{\left(1+2R\left(\Gamma_{uv_{(\alpha)}}\oplus\Gamma_{uv_{(\beta)}}\right)\right)} \tag{1.191}$$

The field equations in the reevaluation of the post-Newtonian approximation is given by the following set of equations, for a linearized theory of gravity,

$$\wp_{0k,k} - \frac{1}{2}\wp_{kk,0} = O\sqrt{\left(1/\left(\Gamma_{uv_{(\alpha)}} \oplus \Gamma_{uv_\beta}\right)/r^5\right)}/1/\left(\Gamma_{uv_{(\alpha)}} \oplus \Gamma_{uv_\beta}\right)$$

$$\wp_{jk,k} - \frac{1}{2}\wp_{,j} = O\sqrt{\left(1/\left(\Gamma_{uv_{(\alpha)}} \oplus \Gamma_{uv_\beta}\right)/r^4\right)}/1/\left(\Gamma_{uv_{(\alpha)}} \oplus \Gamma_{uv_\beta}\right)$$

(1.192)

The newly formed *G.R.F.F.* approximation of the Einstein field equations can be given as

$$8\pi GT^{00} \to \Delta 2\Omega = -2\left(\Gamma_{uv_{(\alpha)}} \oplus \Gamma_{uv_\beta}\right)\delta_{ij}2\left(\Gamma_{uv_{(\alpha)}} \oplus \Gamma_{uv_\beta}\right)2\left(\Gamma_{uv_{(\alpha)}} \oplus \Gamma_{uv_\beta}\right)\delta_{ij}$$

$$-\left(\Gamma_{uv_{(\alpha)}} \oplus \Gamma_{uv_\beta}\right)\delta_{ij,j}2\left(\Gamma_{uv_{(\alpha)}} \oplus \Gamma_{uv_\beta}\right)+2\left(\Gamma_{uv_{(\alpha)}} \oplus \Gamma_{uv_\beta}\right)_{,i}2\left(\Gamma_{uv_{(\alpha)}} \oplus \Gamma_{uv_\beta}\right)_{,i}+2$$

$$\left(\Gamma_{uv_{(\alpha)}} \oplus \Gamma_{uv_\beta}\right)_{,i}2\left(\Gamma_{uv_{(\alpha)}} \oplus \Gamma_{uv_\beta}\right)\delta_{jj,i}+\left[T^{00}+4\left(\Gamma_{uv_{(\alpha)}} \oplus \Gamma_{uv_\beta}\right)T^{00}+T^{ii}\right]$$

(1.193)

$$-\left(\Gamma_{uv_{(\alpha)}} \oplus \Gamma_{uv_\beta}\right)\Delta\left(\Gamma_{uv_{(\alpha)}} \oplus \Gamma_{uv_\beta}\right)+2\Delta\left(\left(\Gamma_{uv_{(\alpha)}} \oplus \Gamma_{uv_\beta}\right)^2\right)-\left(\Gamma_{uv_{(\alpha)}} \oplus \Gamma_{uv_\beta}\right)$$

$$\Delta\left(\Gamma_{uv_{(\alpha)}} \oplus \Gamma_{uv_\beta}\right)=\left(2\left(\Gamma_{uv_{(\alpha)}} \oplus \Gamma_{uv_\beta}\right)-2\left(\Gamma_{uv_{(\alpha)}} \oplus \Gamma_{uv_\beta}\right)^2\right)$$

than in terms of these newly formed functions and parameters we can formulate the original Christoffel symbols as

$$\Gamma^i_{00(2)} \to \left(\Gamma_{uv_{(\alpha)}} \oplus \Gamma_{uv_\beta}\right)$$

$$\Gamma^i_{00(4)} \to \frac{\partial}{\partial x^i}\left(2\left(\Gamma_{uv_{(\alpha)}} \oplus \Gamma_{uv_\beta}\right)^2+\left(\left(\Gamma_{uv_{(\alpha)}} \oplus \Gamma_{uv_\beta}\right)\right)\frac{d^3x'}{|t-t'|}\left[T^{00}\left(x',t\right)+T^{ii}\left(x',t\right)\right]\right)$$

$$+\frac{\partial\varsigma_i}{\partial t}+\frac{\partial^3\chi}{\partial t^2\partial x^i}$$

(1.194)

$$\Gamma^i_{0j} \to -\delta_{ij}\left(\Gamma_{uv_{(\alpha)}} \oplus \Gamma_{uv_\beta}\right)+\frac{1}{2}\left(\frac{\partial\varsigma_i}{\partial x^j}-\frac{\partial\varsigma_j}{\partial x^i}\right)$$

$$\Gamma^0_{00} \to \left(\Gamma_{uv_{(\alpha)}} \oplus \Gamma_{uv_\beta}\right)$$

then in determining the eccentricity, the form of the ellipse, the hyperbola and parabola we refer to the following equations: we begin with the derivative of the area vector

$$dA = \frac{1}{2}\left|r \times d\left(\Gamma_{uv_{(\alpha)}} \oplus \Gamma_{uv_\beta}\right)r\right|$$

(1.195)

and then in terms of the time component

$$\frac{dA}{dt} = \frac{1}{2}\left| r \times \left(\frac{dr}{dt} + \left(\Gamma_{uv_{(\alpha)}} \oplus \Gamma_{uv_\beta} \right) \frac{dr}{dt} \right) \right| = \frac{1}{2}\left| \left(\Gamma_{uv_{(\alpha)}} \oplus \Gamma_{uv_\beta} \right) r \times \left(\Gamma_{uv_{(\alpha)}} \oplus \Gamma_{uv_\beta} \right) r \right|$$

$$= \frac{1}{2}\left(\Gamma_{uv_{(\alpha)}} \oplus \Gamma_{uv_\beta} \right)^2 h = d\left(\frac{r^2\theta}{dt} \right) + \left(\Gamma_{uv_{(\alpha)}} \oplus \Gamma_{uv_\beta} \right)\left(\frac{r^2\theta}{dt} \right) = r\left(\frac{r^2\theta}{dt} \right) + \left(\frac{r^2\theta}{dt} \right)\left(\frac{r^2\theta}{dt} \right)$$

(1.196)

then for the Lagrangian we will have

$$r \times F = L$$

(1.197)

then we will form the following equation

$$L = r \times \left(\Gamma_{uv_{(\alpha)}} \oplus \Gamma_{uv_\beta} \right) mv = \left(r \times r \right)\left(\Gamma_{uv_{(\alpha)}} \oplus \Gamma_{uv_\beta} \right) mv$$

(1.198)

Then in determining the energy law will be beginning with the following equation

$$\left(\Gamma_{uv_{(\alpha)}} \oplus \Gamma_{uv_\beta} \right) r = \frac{dr}{d\theta}\left(\Gamma_{uv_{(\alpha)}} \oplus \Gamma_{uv_\beta} \right) = \frac{dr}{d\theta}\frac{[h]}{r^2} + \frac{dr}{d\theta}\left(\Gamma_{uv_{(\alpha)}} \oplus \Gamma_{uv_\beta} \right)$$

$$= \frac{dr}{du}\frac{du}{d\theta}\frac{[h]}{r^2} + \frac{dr\left(\Gamma_{uv_{(\alpha)}} \oplus \Gamma_{uv_\beta} \right)}{du}\frac{du}{d\theta} = \frac{\left(\Gamma_{uv_{(\alpha)}} \oplus \Gamma_{uv_\beta} \right)}{u^2}\frac{du}{d\theta}\frac{[h]}{r^2}$$

$$= \frac{dr\left(\Gamma_{uv_{(\alpha)}} \oplus \Gamma_{uv_\beta} \right)}{du}\frac{du}{d\theta}$$

(1.199)

BOOK 2:

(G.R.F.F. Approximation in terms of the Newtonian Potential)

1

Basic Concepts: General Definitions of the Gravitational Fields

Introduction

From Kepler's rule, the periodical times of the planets of the orbits that the forces keep the planets in their orbits must be reciprocally as the squares of their distances from the centers about which they revolve.

(i) Newly developed Kepler's law of areas

(ii) The basic relations that govern ordinary orbits described uniformly about the center according to the law of the $G.R.F.F.$

(iii) The $G.R.F.F.$ under which a given focus is described.

(iv) Application to a given body revolving in the circumference of an ellipse under the action of the $G.R.F.F.$ from a given point on the circumference.

(v) An ellipse about a focus will be described under a $G.R.F.F.$ inversely as the square of the distance.

(vi) An ellipse about a focus will be described under an $G.R.F.F.$ inversely as the square of the distance and its application to planetary motion.

(vii) Kepler's third law and its application of the result of planetary motions.

(viii) Given the $G.R.F.F.$'s is inversely as the square of the distance and given its magnitude, we determine the ellipse which the body will describe when projected from a given point with a given velocity in a given direction.

Definition I

A *G.R.F.F.* is that by which bodies are drawn or equally compelled both to and away from a given point as the center of that directed behavior.

This sort of gravity, is by which planets are drawn continually drawn from their original rectilinear motion and made to revolve in curvilinear orbits.

Definition II

A centrifugal force is in fact a rotationally dependent force by which bodies tend towards the center of the Earth, or any orbiting and rotating body. This is a special frame-dependent phenomenon. The difference between a projectile made to revolve in an orbit around the Earth and the Moon made to revolve around the Earth, is by a completely difference force. The Moon is continually drawn from its rectilinear motion by the special *G.R.F.F.* (both by the presence of the Earth and the Sun), from the rectilinear motion by which it would normally move, and made to revolve in the orbit in which it now describes, nor by which the Moon could move without the presence of that special force and retain it in that orbit. The nature of this force has to be just right. If the force were too small the Moon would then not move from its original rectilinear motion, and if it were to great then it would draw the Moon into the Earth directly.

Definition III

The absolute quantity of a *G.R.F.F.* is the measure of the same, proportional to the efficacy of the cause that moves it from the center through the spaces round about.

Definition IV

The accelerative quantity of a rotationally-dependent force is the measure of the same, proportional to the velocity which it generates in a given time.

The force is greater at a less distance and less at a greater distance. This same force is greater in valleys and less on tops of mountains. And also less at greater distances from the body of the Earth; but at equal distances, it is the same everywhere, because it equally accelerates all falling bodies regardless of mass.

Definition V

The motive quantity of a rotationally-dependent force is the measure of the same, proportional to the motion which it generates in a given time.

The weight is greater in a greater body and lesser in a lesser body. It is also greater near to the surface of the Earth and lesser farther away. This quantity is defined as done previously, and is as the body propelled towards the surface of the rotating body, in direct proportion to the magnitude of the rotation of the rotating body. This force, its quantity, labeled as motive, accelerative and absolute forces. They tend to the center and thus the surface of the rotating frame of reference. Then the motive force applied to the body in question as a given form and path and the seemingly natural tendency of the whole towards the surface of the rotating frame of reference, arising from the collection of the parts taken together as a whole; the accelerative force placed on the smaller body as a given power to move all smaller bodies within range of the rotating frame's field of force; and the absolute force to the surface at a given point drawn from the point of the position of the smaller body relative to the surface of the larger body drawn as a straight line to the point of contact on the surface of the larger rotating body. This cause is due to the rotation of the central and larger body in question.

The quantity of downward motion relative to the position of the surface of the rotating body in question, is multiplied by the quantity of matter present, and the motive or nature of the force acting on this smaller body arises from the accelerative force due to the rotating force of the rotating body in question multiplied by the same quantity of matter. The sum of the actions of the accelerative force, placed upon the several parts of the body, is the motive force as a whole. Near the surface of Earth, this frame-dependent force is the same for all bodies, this motive force or the weight of the body being acted upon, is as that body; though if we were to move higher from this surface, this force is proportionally less, this weight would be likewise equally diminished, and would be as the product of the body being acted upon. These forces are referred to as both motive and accelerative.

Corollary *I*

A body acted upon by two equal forces simultaneously, based on the fact they are directed both towards the center of this body being acted upon, and moving in parallel but opposing directions, will have the tendency to move perpendicular to the directions of both of these forces.

If a body were placed at the point A and were impressed upon by two separate forces, denoted F_1 and F_2, will move perpendicular to the directions of these forces and thus begin to move along the straight line AB, looking to end at the point B. If given continuously, directed at the center of this body in question throughout its motion and placed at the right angle, we will lead towards an product of angular rather than linear velocity.

Corollary *II*

The newly derived motion from Corollary I, is obtained by taking the sum of both forces placed at the right acute angle, given arbitrarily, the newly derived motion can be formed.

2

On the notion of limits and ratios of evanescent quantities

Introduction

Here we describe the notions in regard to various different quantities in terms of differential calculus.

Lemma I

Quantities and ratios of quantities in any finite time were to converge continually to an equality and prior to the end of that time will approach nearer to the other given by any difference, will become equal.

Given two quantities $X(t)$ and $Y(t)$, depending continually on time t, and neither of them vanish in the given range $t_0 < t < \infty$, such that $\lim_{t \to t_1} \left[\dfrac{X(t)}{Y(t)} \right] \to 1$

then for some assigned time $t = t_1$ we will have $X(t_1) = Y(t_1)$

But if $X(t_1) \neq Y(t_1)$ will have a contradiction with the prior assumption.

An alternative version of this lemma may be given as if we write the equality such that $X(t) \approx Y(t)$ for $t = t_1$. The for the basic theorem on limits we will have

$X(t) \approx Y(t)$ for $t = t_1 \Leftrightarrow X(t)$ and $Y(t)$ become equal for $t = t_1$

Here the notational device for infinitesimal geometry is given explicitly.

If $X(t)$ and $Y(t)$ tend to zero for $t \to t_1$ we become naturally more careful.

Here the given ratios with which quantities vanish are not truly ratios of ultimate quantities, but limits in which the ratios of quantities decreasing without limit will always converge. They then approach nearer than by any given difference, without going beyond, or attain to, till these quanities are diminished in inifinitum.

Lemmas II, III, IV

Lemma II

Given a figure AacE, terminated by the right lines Aa, Ee and the given curve acE, there will bet the defined parallelograms Ab, Bc, Cd and so on, under the equal bases AB, Bc, Cd and such, and the sides Bb, Cc, Dd, and so on parallel to one side Aa of the figure there will be the parallelograms aKbl, bLcm, cMdn, and such. Then the greater the downward force, due to the rotation of the rotating body, the size of the line aA, bB, cC and so on will be diminished in length in direct proportion to the above mentioned force. The defined parallelograms will likewise decrease in size and area in proportion to the above mentioned parameters.

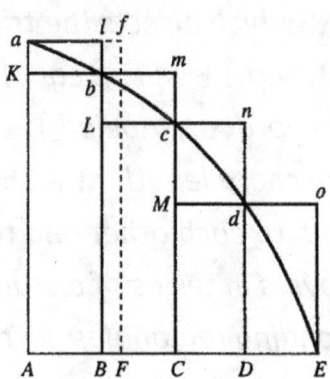

Cor. *I* . The ultimate sum of the defined parallelogarms will be in direct proportion to the defined parameters.

Cor. *II* . The magnitude of the arcs, *ab, bc, cd* and *so* on are in direct proportion to the defined parameters.

Lemma III

If two figures AacE, PprT, are inscribed with two series of parallelograms, equal number in series, and the given two bodies of different masses whose starting point of descension are the same height at respective points a and p, those figures AacE and PprT will be in the same ratio, that is having the same properties.

Lemma IV

Given all homologous sides of each set of inscribed parallelograms, whether the falling mass describes arcs of either curvilinear or rectilinear descent, the properties of those parallelograms will be the same.

Lemmas V-VIII

Lemma V

Given a triangle Adr *, where chords* AD, Dd, AR, *and* Rr *will be equal. Then subtending chords drawn from* Ab *which bisects the triangle* Adr *in half the two parts of the subtending chord,* AB *and* Bb *will be equal. Than another chord drawn directly perpendicular the chord* Ab *, labeled* DR *bisecting it directly in half will be of the same equality of the chord length* Ab *. The two half's of the chord* DR *, labeled* RB *and* BD *, will equal to each other and to* AB *and* Bb *. The two arcs* ACB *and* Acb *, where* Acb *covers in terms of distance along the surface twice that of the other arc the corresponding rectangles next to it, will always have a greater area than the triangles next to the smaller arc regardless of the magnitude of the two existing parameters.*

Lemma VI

If the values of the arc, representing the orbiting motion, are determined by the G.R.F.F.,

than likewise so are the related chord and tangent drawn from the arc. That is given the arc ACB, there is a tangent line AD that is likewise straight, and another chord drawn from A

to B given as the chord AB. The angle formed, its absolute value, given as AB is related directly to the chord and arc already given as determined by the values of the G.R.F.F..

Lemma VII

Given the subtense of the angle of contact in which there is the curve of a orbiting body, than this curvature is as the square of the subtense of that arc that is directly defined, its values of absolute curvature determined by the G.R.F.F..

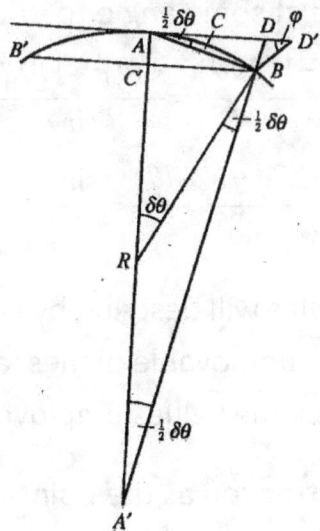

Given the line of force is directed along the singular line $R = AA'$, then the angle of curvature measures the strength of the field. The greater the curvature, the stronger the field. The lines AC and BB' are in direct proportion to each other. The stronger the field then the longer the line AC will then extend, and conversely the shorter line BB' will extend shorter. In terms of angles given in the above figure, we will have the following ratios: The stronger the force, the direct proportional decrease in length of the lines RB and AB'. The following angles are in direct proportion to one another, decrease one then decrease the other. All angles, $\frac{1}{2}\delta\theta$ will decrease in value once the strength of the G.R.F.F. increases. The angle φ will also decrease in magnitude in proportion to the defined force field.

The subtense of the angle of contact at $A = BD$, that is directly perpendicular to the tangent line AD positioned at A. Than the versed sine of the arc BAB is AC, bisecting the chord BB. The versed sine of the arc BAB, its absolute value as the magnitude of curvature, is directly determined by the given G.R.F.F. . This is governed also by the inverse square law excluding the eccentricity of the given orbiting ellipse.

We will then have the following equations

$$AD = 2R\tan 1/2d\theta$$

$$BD = AB\sin 1/2d\theta = \frac{1}{2R}AB^2 + Od\theta$$

Given then that BD is inclined at some angle to AD, then we will also have the relation

$$\frac{AD'}{AB} = \frac{\sin(180 - \varphi - 1/2d\theta)}{\sin\varphi} = \frac{\sin(\varphi + 1/2d\theta)}{\sin\varphi} \to 1$$

$$\to \frac{\sin(180 - (G.R.F.F.) - 1/2d\theta)}{\sin\varphi} = \frac{\sin(\varphi + 1/2d\theta)}{\sin\varphi}$$

The areas which revolving bodies will describe by given radii drawn to a stable center of force, lie in the same unmovable planes, are directly proportional to the times as determined directly by the values employed by the G.R.F.F..

Let us denote the center of attraction as the point S. Then the orbit as determined by the above mentioned G.R.F.F., acting continuously at equal intervals of time δt apart. Then let B, C, D, and so on be the positions of the particles within the field at various arbitrary times.

The dynamics of a particle under the general law of the G.R.F.F.

Given an inertial mass at a distance r, from the centre of attraction is given by the equation

Lemma VIII

The subtense of the angle of contact, in all curves in which at the point of contact and given curvature, is as the square of the subtense of the coterminous arc, all in direct ratio to the magnitude of the G.R.F.F.

Lemma VIX

Given the right lines AR, BR, with the arc ACB, the chord AB and the given tangent AD, form three triangles RAB, RACB, and RAD, and likewise the points A and B approach each other, then the triangles are the same and equal.

Lemma X

If a right line AE, and a given curved line ABC, given both by some position, cut each other at some angle A; then to that right line, in a given angle BD, CE are applied, meetine the curvein B, C; and the points B and C both meet at the point A: then the areas of the curved triangles ABD, ACE, will be equal to each other as the squares of the sides.

3

On the motion of particles under the G.R.F.F.: an introduction to a new treatment

Introduction

Here we focus on the treatment of elliptical motion under the inverse-square law of gravitational motion. We will be focusing on the perturbations of elliptical motion, referred to as the variation of the constants

The dynamics of a particle under the general law of the G.R.F.F.

We begin with the field equations for a generalized *G.R.F.F.* field for associated particles

$$F_{-(ij)} = -mF(r)\frac{r}{r}$$

$$F_{ij} = mF(r)\frac{r}{r}$$

(1.1)

where $\frac{r}{r}$ is an unit vector in the direction of increasing r, and m, the same as the gravitational mass and equal to the inertial mass. We will then have

$$F_T = F_{\alpha\beta} = F_{ij} \oplus F_{-(ij)}$$

(1.2)

and

$$\omega = \frac{d\theta}{dt}$$

(1.3)

then

$$\frac{d\omega}{dt} = \left(F_{ij} \oplus F_{-(ij)} \right) \frac{v}{r^2} dt \frac{r}{r}$$ (1.4)

The conservation of angular momentum

We will then have

$$\frac{d}{dt}(m \times \omega) = \frac{d}{dt} m \times \omega + m \times \frac{d}{dt} \omega$$ (1.5)

The conservation of angular momentum can then be written as

$$[h] = \omega \times m$$ (1.6)

then since by equation (1.6) we will have

$$[h] \cdot m = 0$$ (1.7)

the motion of the particle is restricted to the plane orthogonal to h provided $h \neq 0$
, If

$$[h] = 0$$ (1.8)

we have to move differently. Evaluating the time derivative of r/r we will have

$$\frac{d}{dt} \frac{r}{r} = \frac{(m \times \omega) \times r}{r^3} = \frac{[h] \times r}{r^3}$$ (1.9)

The law of areas

If $h \neq 0$, the orbital plane is normal to h. Choosing the z-axis along the direction of h we will have

$$
\begin{aligned}
[h] &= \left(0, 0, [h]\right) \\
F_i &= \left(F_i \cos\varphi, F_i \sin\varphi\right) \\
F_j &= \left(F_j(-)\cos\varphi, F_j(-)\sin\varphi\right)
\end{aligned}
$$

(1.10)

we will then have

$$
\begin{aligned}
\omega &= \left(F\cos\varphi + (F\sin\varphi), F(-\sin\varphi) + F(-\cos\varphi)\right) \\
&= \left(\frac{d^2 r}{dt^2}\cos\varphi + \left(\frac{d^2 r}{dt^2}\sin\varphi\right), \frac{d^2 r}{dt^2}(-\sin\varphi) + \frac{d^2 r}{dt^2}(-\cos\varphi)\right)
\end{aligned}
$$

(1.11)

and

$$
(m \times \omega) = r^2 \left(F_{ij} \oplus F_{-(ij)}\right) v \frac{z}{z} dt = [h]
$$

(1.12)

or if *A* denotes the area swept by the radius vector in the orbital plane

$$
\frac{dA}{dt} = \frac{1}{2} r^2 \left(F_{ij} \oplus F_{-(ij)}\right) v \, dt = \frac{1}{2}[h]
$$

(1.13)

This is Kepler's law of areas.

The conservation of energy

By the equation

$$
v \cdot \frac{dv}{dt} = \left(\left\{F_{ij}\frac{(r)}{r} \oplus F_{-ij}\frac{(r)}{r}\right\}(v) dt\right) = \left\{F_{ij}\frac{(r)}{r} \oplus F_{-ij}\frac{(r)}{r}\right\}\left(\frac{dr}{dt}\right) dt
$$

(1.14)

therefore

$$\frac{1}{2}\frac{d}{dt}v^2 = \frac{d}{dt}\int_{-\pi}^{\pi}\left(F_{ij} \oplus F_{-(ij)}\right)vdt(r) \tag{1.15}$$

assuming then that $f(r) < r^{-1}$

$$V(r) = \int_{r}^{\infty}\left(F_{ij} \oplus F_{-(ij)}\right)v(r)dt \tag{1.16}$$

then from equation (1.15)

$$\frac{1}{2}v^2 = V(r) + E \tag{1.17}$$

E is a constant. $-V$ is the potential energy and E the total energy. The energy integral can also be given as

$$v^2 + [h]^2 = 2r^2(V + E) \tag{1.18}$$

The equation governing r in the orbital plane

We start with the following equation

$$m\frac{dv}{dt} + \frac{[h]^2}{r^2} + \frac{1}{r^2}(mv^2) = \left(F_{ij} \oplus F_{-(ij)}\right)vdt \times m + \frac{[h]^2}{r^2} + \frac{1}{r^2}(m\cdot\omega)^2 \tag{1.19}$$

we will then have

$$\frac{dv}{dt}(r) = F_{ij}(r)$$
$$\frac{1}{r^2}(m\cdot\omega)^2 \tag{1.20}$$

followed by

$$\frac{d^2r}{dt^2} - \frac{[h]^2}{r^3} = \left(F(r)_{ij} \oplus F(r)_{-ij}\right)vdt / r^2 \tag{1.21}$$

then we can form the following equation

The dynamics of a particle under the inverse-square law of attraction

According to the inverse-square law of attraction we will start with

$$F_{\alpha\beta}(r) = \frac{\mu}{r^2} \tag{1.22}$$

where μ is previously defined. We will then also have

$$\frac{dv}{dt} = \frac{\mu}{r^3}r \tag{1.23}$$

We will as well have

$$V(r) = \int_r^\infty \frac{\mu}{r^2} dr = \frac{\mu}{r} \tag{1.24}$$

and

$$\frac{1}{2}\omega^2 = E + \mu/r \tag{1.25}$$

The Lenz vector and the Lenz equation

As an integral we will, in relation to the inverse-square law, have

$$\mu \frac{d}{dt}\frac{r}{r} = \mu \frac{[h]\times r}{r^3} = \left([h]\times\frac{dv}{dt}\right) \tag{1.26}$$

this equation integrates to give

$$\mu \frac{r}{r} = (v\times[h]) - \mu\{e\} \tag{1.27}$$

here e is a constant and is referred to as the Lenz vector. We can rewrite the equation (1.27) as

$$\mu\left(\{e\}+\frac{r}{r}\right)=v\times[h] \tag{1.28}$$

The scalar product of this equation with h gives

$$e\cdot[h]=0 \tag{1.29}$$

If we choose the origin of φ along e, we will have

$$r\left(1+\{e\}\cos\varphi\right)=[h]^2/\mu \tag{1.30}$$

which is the equation of an ellipse with eccentricity e and semilatus rectum

$$l=\frac{[h]^2}{\mu} \tag{1.31}$$

or

$$l=a(\gamma)\left(1-\{e\}^2\right) \tag{1.32}$$

here a is for the semimajor axis of the ellipse,

$$a=\frac{l}{1-\{e\}^2}=\frac{[h]^2}{\mu\left(1-\{e\}^2\right)} \tag{1.33}$$

The main purpose of the Lenz vector of length e in the orbital plane, is that is ensures the fixity in space of the major axis of the elliptical orbits. The square of the equation (1.27)

$$\mu^2\left(\{e\}^2+\frac{2}{r}\{e\}\cdot r+1\right)=v^2[h]^2 \tag{1.34}$$

through further simplification we will have then

$$u^2\left(\{e^2\}-1\right)=2[h]^2\,E \tag{1.35}$$

since

$$[h]^2 / \mu = a(\gamma)\left(1-\{e\}^2\right) \tag{1.36}$$

we will have

$$a = \left(-\frac{1}{2}\mu / E\right)\gamma \tag{1.37}$$

Kepler's third law

The area of the elliptical orbit is given as

$$\pi a(\gamma)\frac{b}{\gamma} = \pi a^2(\gamma)\sqrt{\left(1-\{e\}^2\right)} \tag{1.38}$$

the constant of the areas is

$$A_{,t} = \frac{1}{2}[h] \tag{1.39}$$

the period of the orbit is given as

$$P = \frac{2\pi}{[h]}a(\gamma)\sqrt{\left(1-(e)^2\right)} \tag{1.40}$$

where

$$[h] = \sqrt{\left[\mu a(\gamma)\left(1-(e)^2\right)\right]} \tag{1.41}$$

then we will have here

$$P = \frac{2\pi}{\sqrt{\mu}}a(\gamma)^{3/2} \tag{1.42}$$

An alternative derivation of the elliptical orbit

From the original Lenz equation, we will have

$$u = \frac{1}{r} = \frac{\mu}{[h]^2}\left(1+(e)\cos\varphi\right) \tag{1.43}$$

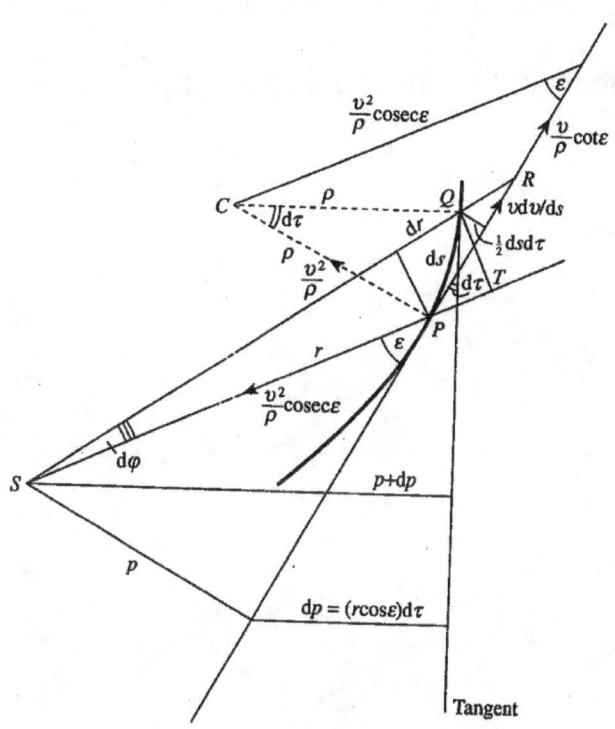

Based on the above figures, we can derive the values, lines and angles and such that one can use to measure the angle of curvature and thus the magnitude of the gravitational source. We begin with the assumption that C is the center of mass and p is the radial line of force form that. S will be the location of the actual source of gravitation. While r is the radial line, in which directs the force tensor to the center point of the line of curvature. The lines p and dp measure the distance bewtween both related gravitation sources. $\frac{v^2}{\rho}$ measures the radial directed line of force as P moves to the point position Q. vdv/ds measures the original linear velocity not under the influence of the gravitational field. Increase the angle of curvature and the strength of the field and the following parameters will change proportaionally: $d\varphi, \varepsilon, ds, dr, d\tau, T, \frac{1}{2}dsd\tau$. With an increase in curvature the following values will decrease in magnitude: $d\varphi, dr, ds, p, p+pd, \varepsilon$. The following

values will increase in value in contrast to the prior values: $d\tau, T, \frac{1}{2}dsd\tau$. $\frac{v^2}{\rho}\mathrm{cosec}\varepsilon$

represents the direction of the radial line of force in one direction while $\frac{v^2}{\rho}\mathrm{sec}\varepsilon$

represents the radial line force moving the direct opposite direction.

4

The Law of areas and some relations that follow

Introduction

The areas by which revolving bodies describe in terms of radii drawn to an immovable center of force, defined as the G.R.F.F., lie in the same immovable planes and are also directly proportional to the related times by which they are described.

Let S be the center of the $G.R.F.F.$. We will now consider the orbit by which is defined in terms of the $G.R.F.F.$, acting as impulses at equal intervals of time δt apart. Then B,C,D,E,F be the positions of the particle in question at times $\delta t, 2\delta, 3t, 4\delta, ect.$

The particle in question originally at A and moving with a defined velocity v, will naturally travel a distance $v\delta t$ in the direction of AB and in the plane containing AV and v, will arrive eventually at B. At this point, under the influence of the $G.R.F.F.$ will then tend to move in the direction of BS and be displaced to the point V. By inertia the particle will naturally move from B to c in the direction AB having travelled a distance $v\delta t = AB$ in the first interval and the same plane. By the nature of the field, the originally defined particle, by composition of the displacements BV and Bc at the end of the second interval $2\delta t$, will find itself at C, where cC is equal and parallel to BV. The areas of $\triangle s SAB$ and SBc will be equal, having equal bases, and the areas of $\triangle s SBc$ and SBC will also be equal.

There fore we will have

$$area\triangle SAB = area\triangle SBC$$

At each point A, C, and so on there will be equal numbers of force vectors positioned at equal angles on both sides of the body in question that will force this body to cut an angle, equally at each said point, directly proportional to the number of force vectors involved. The areas described in two successive intervals of time, an equal amount apart, are themselves equal, as the body continues to move in the plane SAB. We then have

$$area \triangle SBC = area \triangle SCD$$

The moving body describes the same area during the third interval δt as each of the first two intervals. In terms of induction, equal areas are described in all intervals of time δt apart.

The given triangle will be augmented and length diminished in infinity, the perimeter ADF will be a curved line and by the G.R.F.F. the body is taken continually from the tangent of this curve, will always act continually, and areas such as $SADS$ and $SAFS$ will be described such that they are always proportional to the times.

Corollary I

We will first have

$$AB = v_A \delta t = \left(F_{ij} \oplus F_{-(ij)} \right) \delta t$$
$$BC = v_B \delta t = \left(F_{ij} \oplus F_{-(ij)} \right) \delta t \qquad (1.44)$$
$$CD = v_C \delta t = \left(F_{ij} \oplus F_{-(ij)} \right) \delta t$$

the following areas are equal SAB, SCB, SCD, ect. such that

$$p_A AB = p_B BC = p_C CD = ect. \qquad (1.45)$$

where p_A, p_B, p_C are the lengths of the perpendiculars from s to AB, BC, CD ect. We obtain then

$$v_A p_A = v_B p_B = v_C p_C = ect. \tag{1.46}$$

Corollaries II

The displacements BV and EZ, that tend to the versed sines of the arcs AC and DF as $\delta t \to 0$, are caused by the impulses generated by the G.R.F.F. acting intermittently at B and D, in the limit $\delta t \to 0$, when the afore mentioned force acts continuously.

In spaces devoid of resistance, are drawn away from the original rectilinear motion turned into the curvilinear orbits are to each other as the versed sines of the arcs described in equal times. The greater the G.R.F.F., the proportionally greater the versed sines of the arcs.

Corollary III

Every body that moves in a curved line described in a plane, and by a radius drawn to a point either stationary or moving in a straight line, and another radius drawn on top of the original radius away from that same point, then describes about that original point areas proportional to the times, and is urged by the G.R.F.F. both towards and away from that point.

In that planar orbit, equal times described in equal areas, that orbit is described under the influence of the G.R.F.F.. The following areas are equal SAB, SBC, SDC, SDE, ect. deducing the G.R.F.F. nature of that field. If we extend AB by an equal length c, then according to the laws of inertia the same body will have arrived at c; and the areas of SAB and SBC are equal. The areas SBc and SBC are equal too. Then cC must be parallel to CB, and the force at cC draws the body from Bc to BC then acts in the opposing radial directions in the limit $C \to B$. Likewise the G.R.F.F. dD, that draws the body from Cc to CD, acts in the opposing radial directions CS.

Corollaries IV

If in equal times unequal areas are described, we assume that the body in question is acted on by forces in addition to the $G.R.F.F.$. If then the areas described in equal times continually increase or decrease, then the additional force acting on the body in question must either accelerate or decelerate in the direction of its motion. This occurs because another body blocks the incoming or outgoing radiation on the initial body in question that affects its original orbital motion.

Proposition I. Theorem I

A body, by a radius drawn to the centre of another body, however moved, describes areas about that centre proportional to the times is moved by a force, deemed the G.R.F.F., tending to that other body and of all the accelerative force by which the other body is moved.

Given body L moves about body T, in equal areas and in equal times in a planar orbit, while T is subject to the subtraction of certain forces within the $G.R.F.F.$ field, due to the motion of another body, then L is subject to the total action of the $G.R.F.F.$ towards and away from T, including the subtraction of certain forces due to the presence of another body.

We assume that L is larger than T. Then as T rotates around L in an orbital path, L will gyrate on its axis in relation to T. The larger the difference in mass of these two bodies the greater the orbital radius and the diameter of the path of axis gyration. The force acting on L is the . due to the presence of the larger body T.

Corollaries V

If unequal areas are described by L, relative to T, in equal times, then the forces acting L and T is due to the combination of the generated G.R.F.F. fields produced by the presence of both bodies.

Proposition II. Theorem II

The G.R.F.F. 's of bodies, which describe different circles, tend towards and away simultaneously from the centre's of the same circles; and are to each other as the squares of the arcs described in equal times divided by the radii of the circles. The greater the radii then the greater the squares of the arcs. The total angle of the arcs defined by the orbiting body in question is directly proportional to the number of the lines of force defined by the G.R.F.F.

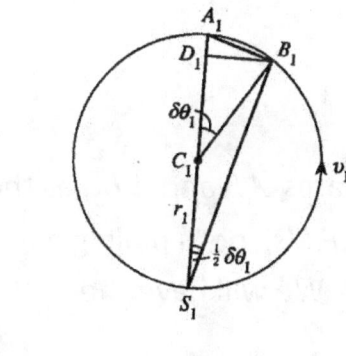
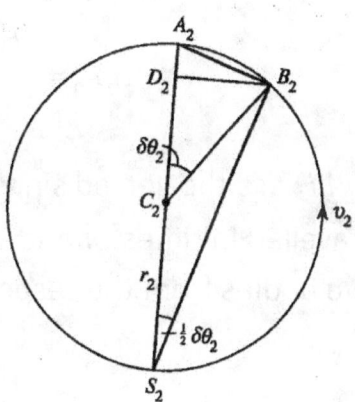

Given two circles or radii r_1 and r_2, described uniformly with two velocities v_1 and v_2 and the arcs $A_1B_1 = r_1\delta\theta_1$ and $A_2B_2 = r_2\delta\theta_2$ are described in times δt_1 and δt_2, we will then have

$$arc A_1 B_1 = r_1 \delta\theta_1 = v_1 \delta t_1 = \left(F_{ij} \oplus F_{-(ij)} \right) \frac{mM}{r^2} \delta t_1 \tag{1.47}$$

and

$$arc A_2 B_2 = r_2 \delta\theta_2 = v_2 \delta t_2 = \left(F_{ij} \oplus F_{-(ij)} \right) \frac{mM}{r^2} \delta t_2 \tag{1.48}$$

then by elementary geometry

$$A_1 S_1 \times A_1 D_1 = \left(A_1 B_1 \right)^2 \tag{1.49}$$

and

$$A_1 S_1 \times A_1 D_1 = \left(A_1 B_1 \right)^2 \tag{1.50}$$

it follows then that

$$A_1 D_1 = \frac{\left(F_{ij} \oplus F_{-(ij)} \right) \frac{mM}{r^2} \delta t_1}{2r_1}$$

$$A_2 D_2 = \frac{\left(F_{ij} \oplus F_{-(ij)} \right) \frac{mM}{r^2} \delta t_2}{2r_2} \tag{1.51}$$

Since $A_1 D_1$ and $A_2 D_2$ are the versed sines of twice the arcs $A_1 B_1$ and $A_2 B_2$, as they are the distance travelled in times δt_1 and δt_2 by the G.R.F.F's, both pulling and pushing the two bodies from the respective centres. We will have then

$$\frac{1}{2}\left(G.R.F.F. \right)_1 = \frac{A_1 D_1}{\left(\delta t_1 \right)^2} = \frac{1}{2r_1} = \frac{\left(\left(F_{ij} \oplus F_{-(ij)} \right) \frac{mM}{r^2} \right)^2}{\left(\delta t_1 \right)^2}$$

$$\frac{1}{2}\left(G.R.F.F. \right)_2 = \frac{A_2 D_2}{\left(\delta t_2 \right)^2} = \frac{1}{2r_1} = \frac{\left(\left(F_{ij} \oplus F_{-(ij)} \right) \frac{mM}{r^2} \right)^2}{\left(\delta t_2 \right)^2} \tag{1.52}$$

and if $\delta t_1 = \delta t_2$ then

$$\frac{(G.R.F.F.)_1}{(G.R.F.F.)_2} = \frac{\left(\left(F_{ij} \oplus F_{-(ij)}\right)\frac{mM}{r^2}\right)_1^2}{r} = \frac{\left(\left(F_{ij} \oplus F_{-(ij)}\right)\frac{mM}{r^2}\right)_2^2}{r_2} \tag{1.53}$$

Corollary I

By using equation (1.47), equations (1.52) gives

$$(G.R.F.F.)_1 : (G.R.F.F.)_2 = \left(\left(F_{ij} \oplus F_{-(ij)}\right)\frac{mM}{r^2}\right)_1^2 / r_1 = \left(\left(F_{ij} \oplus F_{-(ij)}\right)\frac{mM}{r^2}\right)_2^2 / r_2 \tag{1.54}$$

Corollary II

If the periodic times in which the orbits are described are T_1 and T_2, then

$$T_1 : T_2 = r_1 / \left[\left(F_{ij} \oplus F_{-(ij)}\right)\frac{mM}{r^2}\right]_1 : r_2 / \left[\left(F_{ij} \oplus F_{-(ij)}\right)\frac{mM}{r^2}\right]_2 \tag{1.55}$$

or

$$\left[\left(F_{ij} \oplus F_{-(ij)}\right)\frac{mM}{r^2}\right]_1 : \left[\left(F_{ij} \oplus F_{-(ij)}\right)\frac{mM}{r^2}\right]_2 = r_1 / T_1 : r_2 / T_2 \tag{1.56}$$

Corollary III

If $T_1 = T_2$ then

$$T_1 : T_2 = \left[\left(F_{ij} \oplus F_{-(ij)}\right)\frac{mM}{r^2}\right]_1 : \left[\left(F_{ij} \oplus F_{-(ij)}\right)\frac{mM}{r^2}\right]_2 \tag{1.57}$$

Corollary IV

The relations (1.47) and (1.53) then

$$arc = vt = \left[\left(F_{ij} \oplus F_{-(ij)} \right) \frac{mM}{r^2} \right] \tag{1.58}$$

and

$$G.R.F.F. = \left[\left(F_{ij} \oplus F_{-(ij)} \right) \frac{mM}{r^2} \right]^2 / r \tag{1.59}$$

the angle described in a time t is in the force equations.

Proposition *III* . Problem *I* .

Being given, in any places, the velocity of a body which describes a given figure, by means of the G.R.F.F. *'s, are related to some common centre: to find that centre.*

Let PT , TQV , and VR be the directions of the velocities at three points P, Q, and R along the given orbit so that these are likewise the directions of the tangents to the orbit at these points. We seek to find the location of the centre of the G.R.F.F., S, under which the orbit is described.

Let PA, QB and RC be normal to the tangents at P, Q, and R and the lengths inversely proportional to the angular velocities v_P, v_Q and v_R at these points that is

$$PA : QB : RC = \left[\left(F_{ij} \oplus F_{-(ij)} \right) \frac{mM}{r^2} \right]^{-1}_P : \left[\left(F_{ij} \oplus F_{-(ij)} \right) \frac{mM}{r^2} \right]^{-1}_Q + \left[\left(F_{ij} \oplus F_{-(ij)} \right) \frac{mM}{r^2} \right]^{-1}_R \tag{1.60}$$

If S is the centre of attraction, we can drop the perpendiculars SP and ST to the tangents at R, Q and P. We then draw CE, EBD and DA parallel to the tangents at

R, Q and P intersecting at the points E and D. We then drop the perpendiculars DP'' and DT'' to the tangents at P

and Q. We will then have

$$\frac{SP'}{ST'} = \frac{\left[\left(F_{ij} \oplus F_{-(ij)}\right)\frac{mM}{r^2}\right]_P}{\left[\left(F_{ij} \oplus F_{-(ij)}\right)\frac{mM}{r^2}\right]_P} = \frac{AP}{BQ} = \frac{DP''}{DT''} \qquad (1.61)$$

Newton's relations replaced with the G.R.F.F. for determining the law of attraction from the orbit

Proposition IV. Theorem III

In a space void of resistance, if a body revolves in any orbit about an immovable center, and in the least time describes an arc just then nascent and the versed sine of the arc is drawn then bisecting the chord and passing through the centre of force, the force in the middle of the arc will directly as the versed sine and inversely as the square of the time. The angle and sine of the said angle is directly proportional to the magnitude of the G.R.F.F. and inversely proportional to square of the distance between the two gravitating bodies. Each force vector described on the rotating body in question, regardless of the value of r^2, will cut equal angles as the lines are drawn through the body in question. The total number of force vectors described, determined proportional to r^2, determines the value of the arc, that is its versed sine.

Let be the orbit described by a particle about the centre of attraction S; P and Q the neighboring positions of the particle at an interval δt apart. YPZ, the tangent at P specifying the direction of motion, and QR the continuation of SQ, is the versed sine of twice the arc PQ. We then let SY be the perpendicular to the tangent at P.

In the absence of the *G.R.F.F.*, the body, originally at *P*, will move by inertia in a straight line and arrive at R in the time interval δt tangent at *P*. By the *G.R.F.F.* acting for the time interval δt the body is drawn to the point *Q*.

Corollary I

By proposition *I* the area of the triangle *SPQ* is

$$G.R.F.F. = \left(\frac{1}{2}[h]\right)\delta t \tag{1.62}$$

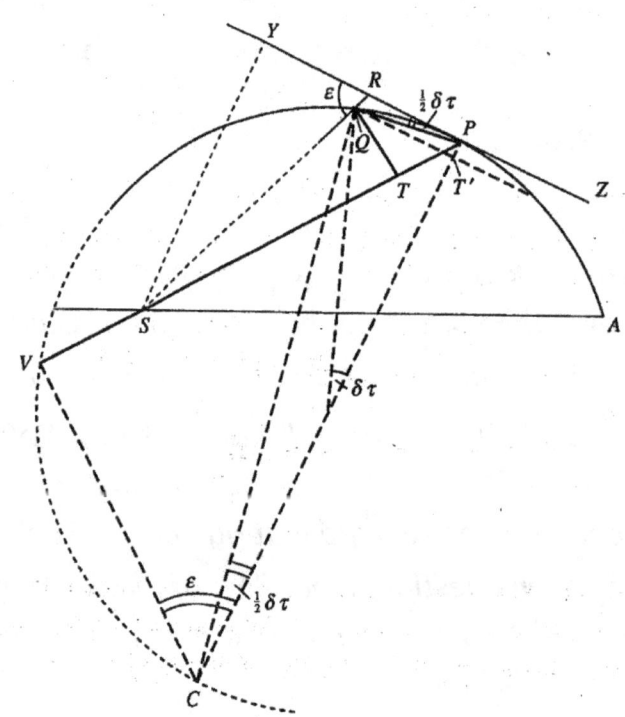

Here in this diagram *S* represents the central source of gravitation. *P* represents the position at a singular instant of the orbiting body. The line *YPZ* represents the tangent line to the line of curvature. *C* represents the point of curvature for an ordinary circle of revolution. Increase the angle of curvature of the orbiting body and the following values: measurements of angles themselves, will decrease proportionally, $\varepsilon, \frac{1}{2}\delta\tau, \delta\tau$. The following values, lines and angles themselves are in direct proportion to the magnitude of the curvature: $QT, T'QT'$. In contrast the following values will increase in proportion to an increase in curvature: $\varepsilon, \frac{1}{2}\delta\tau$, where here $\frac{1}{2}\delta\tau$ is the value places between the following lines QP and RP.

We can redraw the diagrams of this proposition to point towards that at P the center of the circle C, passing through P and Q and touching YP at P is on PC perpendicular to YP; in the limit $Q \to P$, the circle becomes the circle of contact.

$$QP^2 = 2\rho PT' = 2\frac{\rho_r}{F(r)}PT' \tag{1.63}$$

Corollary II

Since the velocity *v*, along the orbit is given we will have then

$$G.R.F.F. = \frac{\left(F_{ij} \oplus F_{-(ij)}\right)\frac{mM}{r^2}}{PV} = \frac{\left(F_{ij} \oplus F_{-(ij)}\right)\frac{mM}{r^2}}{2\frac{\rho}{F(r)}\sin\varepsilon} \tag{1.64}$$

Proposition V. Problem II

If a body moves in the semicircumference PQA; we seek to find the law of the $G.R.F.F.$ tending to and away from a point s, so that all lines PS, RS, may be taken for parallels.

By the similarity of $\triangle s$ CPM, PZT and RZQ we have the following relations

$$\frac{CP}{PM} = \frac{PZ}{ZT} = \frac{RZ}{ZQ} = \frac{PZ - RZ}{ZT - ZQ} = \frac{RP}{QT} \tag{1.65}$$

If PC is to r_1 and LP is to r_{-1} then we draw the following relations. The stronger the force of $G.R.F.F.$, then the following decrease in the lengths of the following drawn lines. PRZ is the tangent line to the circle APU, so that

$$(PC \cdot LC)\frac{v}{r^2} = \omega \tag{1.66}$$

We draw the following dependent relations:

$$\frac{PC}{GRFF} = \frac{LC}{GRFF} : \frac{PT}{GRFF} = \frac{LT}{GRFF} : \frac{PM}{GRFF} = \frac{LM}{GRFF}$$
$$= \frac{MS}{GRFF} : \frac{RQ}{GRFF} = \frac{LR}{GRFF} : \frac{QN}{GRFF} = \frac{LRQ}{GRFF} : \frac{NS}{GRFF} = \frac{NR}{GRFF} \tag{1.67}$$

The semidiameter of the circle, AC is directly proportional to the strength of $G.R.F.F.$. AC is directly equal to PC. The angle $\angle RPT$ cut by the tangent PRZ is directly proportional to the magnitude of AC and is equal to the radius of the circle, or inversely proportional to the square of PC and AC. If the rotating body is positioned at the point P, it will naturally move along the tangent PRZ but under the influence of the $G.R.F.F.$ will eventually move to A. The length of the arc PQA is directly proportional to the magnitude of the $G.R.F.F.$. The following lines will increase with the strength of $G.R.F.F.$ increasing as well:

$AN, AM, ANM, AC, PC, MP, TP, MTP$, angle AQ, and angle PQ and angle AQP. The line QT cutting at point Q, and forming parallelogram $NQTM$ directly perpendicular to PTM will be increased in likewise ratio.

Illustrations of the basic known relation

Proposition VI. Problem III

If a body revolves in a spiral PQS, cutting all radii Sp, SQ ect., in a given angle: we seek to find the law of the $G.R.F.F.$ tending toward and away from the center of that spiral.

The radius vector, drawn from a fixed point to any point on the curve, makes a constant angle to the tangent at that points which we term as the equinangular spiral. We will then have

$$G.R.F.F. = P(v) = \left[h^2 \right] v + \left[h \right]^2 \left(\frac{d^2 r_1}{dt^2} \cdot \frac{d^2 r_{-1}}{dt^2} \right) m = \left(\left[h \right]^2 \cos ec^2 \right) r^3 = \frac{\left[h \right]^2}{r^3 \sin^2 \epsilon} \qquad (1.68)$$

The radius of curvature along a given curve $r(\varphi)$, in plane polar coordinates is

$$\frac{1}{\rho} = \left\{ v + \left(\frac{d^2 r_1}{dt^2} \cdot \frac{d^2 r_{-1}}{dt^2} \right) m \right\} \sin^3 \epsilon \qquad (1.69)$$

for an equiangular spiral this will give us

$$\frac{1}{pr} = \sin \epsilon : \frac{1}{\rho r \sin \epsilon} = 1 \qquad (1.70)$$

we will then have

$$G.R.F.F. = \frac{1}{2} r^2 \left\{ v + \left(\frac{d^2 r_1}{dt^2} \cdot \frac{d^2 r_{-1}}{dt^2} \right) m \right\} \qquad (1.71)$$

For the spiral PQS all angles, related to ε, whose magnitude are directly proportional to the strength of the $G.R.F.F.$. The lengths QT and $Q'T'$ are likewise equal to each other and proportional to the strength of the $G.R.F.F.$. The following angle is also proportional, $\delta\varphi$. We can measure the magnitude of the tending forces causing the spiral motion by measuring the magnitude of the following two angles, $\varepsilon, \delta\varphi$ that decrease with an increase in the strength of the field in question. The following lines also decrease in length with an increase in field strength; SQ', SP', SQ, SP.

Proposition *VII* . Problem *IV*

If a body revolves in an ellipse; we seek to find the law of G.R.F.F. *tending both to and away from the center of the ellipse.*

Given two close points P and Q on the ellipse (with C as the centre and semi axis CA and CB) we draw lines QT and QR perpendicular and parallel to CP. Let DCK be the diameter conjugate to PCG and parallel to PR, the tangent at P. We draw PF perpendicular to DCK and QV parallel to Rp. If AC and CB are the following semimajor and major axis, according to the frame-dependent classification we will have $CA(\gamma):CB/\gamma$.

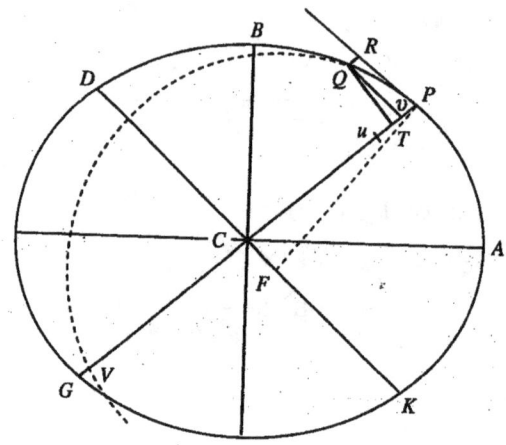

$$\frac{CB}{G.R.F.F.} = \frac{CQ}{G.R.F.F.} : \frac{CA}{G.R.F.F.} = \frac{CL}{G.R.F.F.}$$

$$: \frac{CD}{G.R.F.F.} = \frac{CP}{G.R.F.F.} = \frac{CG}{G.R.F.F.} = \frac{CK}{G.R.F.F.}$$

(1.72)

The total area of the parallelogram, $QRPv$, is directly proportional to the values of $G.R.F.F.$. The angle formed at the corner of v, P, Q and R are also directly proportional to $G.R.F.F.$.

By the property of the ellipse we will have

$$\frac{P\left(\left(F_{ij}\oplus F_{-(ij)}\right)\frac{mM}{r^2}\right)\cdot\left(\left(F_{ij}\oplus F_{-(ij)}\right)\frac{mM}{r^2}\right)G}{Q\left(\left(F_{ij}\oplus F_{-(ij)}\right)\frac{mM}{r^2}\right)^2} = \frac{CP^2}{CD^2}$$

(1.73)

by the similarity of the $\Delta s QvT$ and PCF

$$\frac{Q\left(\left(F_{ij} \oplus F_{-(ij)}\right)\frac{mM}{r^2}\right)}{QT^2} = \frac{CP^2}{PF^2} \tag{1.74}$$

also

$$\frac{QR}{P\left(\left(F_{ij} \oplus F_{-(ij)}\right)\frac{mM}{r^2}\right)} = 1 \tag{1.75}$$

then since $QRPv$ is a parallelogram, we then multiply these above equations to give

$$\frac{CP^4}{PF^2 \cdot CD^2} = \frac{CP^4}{CA^2 \cdot CB^2} \rightarrow \frac{QR \cdot \left(\left(F_{ij} \oplus F_{-(ij)}\right)\frac{mM}{r^2}\right)G}{QT^2} \tag{1.76}$$

Now we associate v with a point r on the left, equidistant from T so that

$$Tr = Tv = T\left(\left(F_{ij} \oplus F_{-(ij)}\right)\frac{mM}{r^2}\right) \tag{1.77}$$

and to define another point V by the following condition

$$rV = \left(Q\left(\left(F_{ij} \oplus F_{-(ij)}\right)\frac{mM}{r^2}\right)\right)^2 / P\left(\left(F_{ij} \oplus F_{-(ij)}\right)\frac{mM}{r^2}\right) \tag{1.78}$$

or by equation $\cot \in = \dfrac{-dr}{rd\varphi}$ we will have

$$\frac{CD^2}{CP^2} \rightarrow \frac{rV}{\left(\left(F_{ij} \oplus F_{-(ij)}\right)\frac{mM}{r^2}\right)G} = \frac{\left(Q\left(\left(F_{ij} \oplus F_{-(ij)}\right)\frac{mM}{r^2}\right)\right)^2}{P\left(\left(F_{ij} \oplus F_{-(ij)}\right)\frac{mM}{r^2}\right) \cdot \left(\left(F_{ij} \oplus F_{-(ij)}\right)\frac{mM}{r^2}\right)G} \tag{1.79}$$

by equation (1.74) we will then have

$$\left(Q\left(\left(F_{ij}\oplus F_{-(ij)}\right)\frac{mM}{r^2}\right)\right)^2 + Pr\cdot P\left(\left(F_{ij}\oplus F_{-(ij)}\right)\frac{m}{r^2}\right)=\left(Q\left(\left(F_{ij}\oplus F_{-(ij)}\right)\frac{mM}{r^2}\right)\right)^2$$

$$+\left(PT+Tr\right)\left(PT-T\left(\left(F_{ij}\oplus F_{-(ij)}\right)\frac{mM}{r^2}\right)\right)=\left(Q\left(\left(F_{ij}\oplus F_{-(ij)}\right)\frac{mM}{r^2}\right)\right)^2 \qquad (1.80)$$

$$+PT^2 -T\left(\left(F_{ij}\oplus F_{-(ij)}\right)\frac{mM}{r^2}\right)^2$$

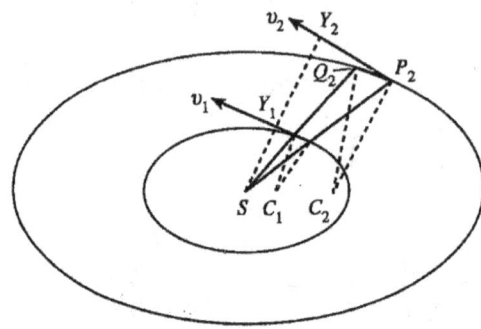

Cor. I. The $G.R.F.F.$ is as the distance of the body from the center of the ellipse and vice versa; if the force is as the distance, the body will move in an ellipse whose center coincides with the center of force.

a. The following is the case of similar ellipses.

(i) $G.R.F.F.=\left(F_{ij}\oplus F_{-(ij)}\right)\dfrac{v}{\rho}\cos ec \in$

(ii) $G.R.F.F._{\cdot 1}:G.R.F.F._{\cdot 2}=\left(F_{ij}\oplus F_{-(ij)}\right)\dfrac{v_1}{\rho_1}:\left(F_{ij}\oplus F_{-(ij)}\right)\dfrac{v_2}{\rho_2}$

(iii) (area of ellipse)/(constant of areas) =

$$T_1:T_2 = \frac{a_1\left(\gamma\right)b_1/\gamma}{p_1\left(F_{ij}\oplus F_{-(ij)}\right)\dfrac{v_1}{\rho_1}}=\frac{a_2\left(\gamma\right)b_2/\gamma}{p_2\left(F_{ij}\oplus F_{-(ij)}\right)\dfrac{v_2}{\rho_2}}=1:1$$

b. The case of two ellipses with semiaxes (a,b) we will have

(i) $G.R.F.F._{\cdot B_1}=G.R.F.F._{\cdot B_2}=\left(F_{ij}\oplus F_{-(ij)}\right)\dfrac{v_1}{\rho_1}:\left(F_{ij}\oplus F_{-(ij)}\right)\dfrac{v_2}{\rho_2}$

(ii) $\rho_1 : \rho_2 = \left(a_1^{\,2}(\gamma)/b_1/\gamma \right) : \left(a_2^{\,2}(\gamma)/b_2/\gamma \right)$

5

The motion of bodies along conic sections

Introduction

Here we focus on the inverse-square law of attraction deduced for bodies revolving in ellipses under the action of the G.R.F.F., both towards and away from the focus of the ellipse at the same time.

Proposition *VIII*. Theorem *I*

If a body revolves in an ellipse, we seek to find the law of the G.R.F.F. *tending both away and towards of the focus of the ellipse.*

Description of figure

S and H are the foci and C is the centre of the ellipse:

CA and CB are the semimajor and semiminor axes of lengths a and b ;

(these are frame-dependent quantities; given as $a(\gamma):\dfrac{b}{\gamma}$);

P and Q are the neighboring points $(Q \rightarrow P)$;

DCK is the diameter, conjugate to PCG and parallel to RPZ ;

PF is perpendicular to DCK and CZ is parallel to SP ;

QR and QT are parallel and perpendicular to SP ;

$Q\left(\left(F_{ij}\oplus F_{-(ij)}\right)\dfrac{mM}{r^2}\right)$ and IH are parallel to RPZ and DCK ;

$\left(\left(F_{ij}\oplus F_{-(ij)}\right)\dfrac{mM}{r^2}\right)$ is on PC and x is PS .

Some properties of the ellipse needed in the solution

(i) $(SP+PH)\gamma = a(\gamma)$

(ii) The latus rectum is twice the semilatus rectum, given as

$$L = 2\frac{b^2}{\gamma}/a(\gamma) = 2\frac{BC^2}{\gamma}/CA(\gamma)$$

(iii) $\angle IPR = \angle HPZ : \angle PIH = \angle PHI : PI = PH$

(iv) $Pv\cdot vG/(Qv)^2 = CP^2/CD^2$

(v) $PF^2\cdot CD^2 = CB^2/\gamma\cdot CA^2(\gamma)$

A simple consequence of the above relations

Proposition *IX*. Theorem *II*

If a body is to move in a hyperbola; we seek to find the law of the G.R.F.F. *tending both simultaneously towards and away from the focus of that figure.*

We find out the force tending from and towards the centre C of the hyperbola. This will be proportional to the distance CP .

The motion of a body along a parabola

Lemma I

The latus rectum of a parabola belonging to any vertex is four times the distance of that vertex from the focus of the figure.

Lemma II

The perpendicular, let fall from the focus of that said parabola on its tangent, is in proportion between the distances of that focus from the point of contact, and the principle vertex of that figure.

S is the focus and A is the vertex of the parabola, such that

$$y^2 = 4a(\gamma)x \qquad (1.81)$$

the origin of the coordinate system at A.

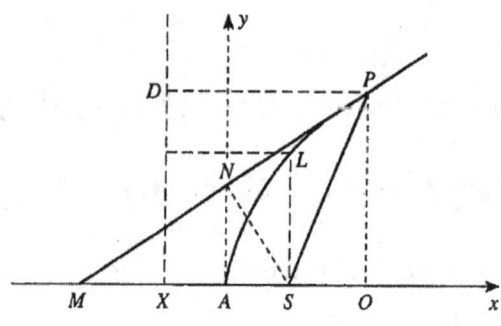

DX is the directrix; P is a random point on the parabola and PD is drawn perpendicular to SX. Then

$$SP = PD = OX \qquad (1.82)$$

is the tangent at P; and SN is perpendicular to PM. The greater the force then the greater the curvature of the arc ALP. Increase the arc then the following things will change in direct proportion to the change in curvature of the arc. The lines *PO* and *LS* will decrease in length while the lines *NS* and *DP* will increase in length. The drawn parallelogram *ANLS* will decrease in area.

We then have

$$LS(\gamma) = SX(\gamma) = 2a(\gamma) \tag{1.83}$$

where

$$L = 4a(\gamma)$$

$2a$ is the semilatus rectum. The parametric formation of the parabola is given as

$$x = a(\gamma)f(r)^2 : y = 2a(\gamma)f(r) \tag{1.84}$$

then

$$P = (AO, OP) = \left(a(\gamma)f(r)^2, 2a(\gamma)f(r)\right) \tag{1.85}$$

following that we will have

$$SO = AO - AS = a(\gamma)\left(f(r)^2 - 1\right) \tag{1.86}$$

the equation of the tangent at *P* is

$$y = \frac{x}{f(r)} + a(\gamma)f(r) \tag{1.87}$$

when

$$y = 0, x = -a(\gamma)f(r)^2 \tag{1.88}$$

therefore

$$AM = AO = a(\gamma)f(r)^2 \tag{1.89}$$

we will then by (v) have

$$SP^2 = SO^2 + OP^2 = \left[a(\gamma)\left(f(r)^2 - 1\right)\right]^2 + 4a(\gamma)^2 f(r)^2 = \left[a(\gamma)\left(f(r)^2 + 1\right)\right]^2 \quad (1.90)$$

Hence

$$SP = a(\gamma)\left(f(r)^2 + 1\right) = AM + AS = SM \quad (1.91)$$

by this relation we will have $MN = NP$, since SN is perpendicular to MP. And since $AO = MA$, AN is parallel to PO and perpendicular to OM. And since $\triangle s\, SAN$ and SNP are similar we will have

$$\frac{SN}{SA} = \frac{SP}{SN} \quad (1.92)$$

or

$$SN^2 = SP \cdot SA = a(\gamma)SP \quad (1.93)$$

SN is the mean proportional of the distances SP and SA. It follows then that

$$\frac{SN^2}{SP} = SA = a(\gamma) \quad (1.94)$$

Proposition X. Theorem III

If a body moves in the parabola; we seek to find the law of the G.R.F.F., tending towards and away from the focus of that figure.

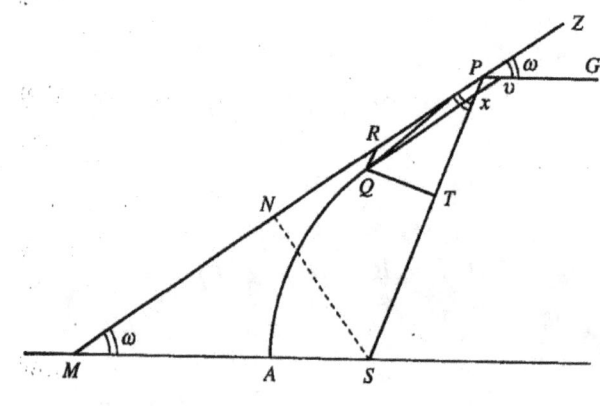

S is the focus and A is the vertex of the parabola;

P and Q are neighboring points on the parabola;

RPZ is the tangent at P;

SN is perpendicular to MRP;

RQ and QT are parallel and perpendicular to SP; PG is parallel to MS.

We will then have the following proportions:

The greater the force and thus the angle of curvature; the following lines will decrease in proportional length: QT and Tx. The following angle will also decrease in magnitude by the same values: ω.

We will then have

$$\angle GPZ = \angle Pvx = \angle SPM = \angle Pxv = \omega \tag{1.95}$$

Where

$$\omega = \left(\left(F_{(ij)_\alpha} \oplus F_{(ij)_\beta} \right) \frac{mM}{r^2} \right) \tag{1.96}$$

The angle ω is related to the parameter μ such that for the tangent at P we will have

$$\cot \omega = \mu = \left(F_{(ij)_\alpha} \oplus F_{(ij)_\beta} \right) \tag{1.97}$$

The equation of the parabola with respect to the oblique axes PG (the x-axis) and PM

(the y-axis) is given by (since $x = Pv : y = Qv$):

$$y^2 = 4\left(a(\gamma) \cos ec^2 \omega \right) x = 4a(\gamma)\left(1 + \mu^2 \right) X \tag{1.98}$$

Which is also given as

$$y^2 = 4SPx \tag{1.99}$$

The stronger the force the greater the curvature of the parabola.

Depending directly on the angle of the lines of force upon the moving body, along the parabolic path, and the relative total number of lines of force on opposing sides the following values will change proportionally: PTS will be decreased proportionally; QT will be diminished; angle ω drawn from the tangent line $MNRPZ$ will be decreased; SN will be smaller; RQ will be diminished; the line PTS will be made smaller; and the angle ω_2 drawn from $\angle GPZ$ and $\angle TPR$ will be increased proportionally.

In the oblique system of coordinates, we will have

$$P=(0,0):Q=(Pv,Qv) \tag{1.100}$$

then by equation

$$Pu \cdot PV = PQ^2$$

Cor. I . From the last Propositions it follows then that, any given body P, goes from the place P with a given velocity in the direction of a right line PR, at the same given time is moved by the *G.R.F.F.* that is inversely proportional to the square of the distance of the places from the given center, this body will move in the given conic section, with its focus in the center of that field of force. The focus, the point of contact as well, the tangent given likewise, a conic section will be described with a given curvature.

Here in the below figure we form the following ratios: given the following arc PQ with tangent line PR and moving linear velocity v, we then will have the following line, decreased in length in direct proportion to increase in curvature of the arc, TQ. The following line will increase in length, denoted as QR. The following angle of curvature will decrease in magnitude with in an increase in force, ε.

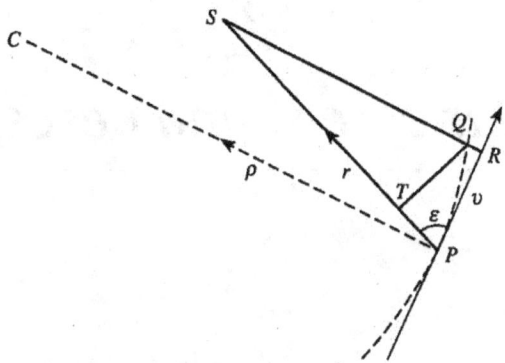

Let *S* be the center of attraction, of the gravitational field, with some known force and constant: then given the position of the particle a given distance away *r*, from *S* ad *PR*, inclined at an angle to *SP* with a given velocity *v*. At an instant of time later, by inertia, the particle will at a given position R, at a distance away from *P*. Due to the *G.R.F.F.* the particle will be found at *Q* along *SR*. We can draw the circle of contact passing through *P* and *Q*, and let p be the radius of curvature. The normal acceleration towards and away from the center is determined by the G.R.F.F.

Kepler's third law

Proposition *XI* . Theorem *IV*

Given right lines being drawn to the bodies that touch orbits, and their perpendiculars let fall on those given tangents from the knonw common focus: the velocities of the bodies vary inversely as the perpendiculars and directly as the square roots of the principal later recta.

6

The rectilinear ascent and descent of bodies

Introduction

By experiments, the descent of bodies is varied as the square of the time and the motion of projectiles was in the curve of a parabola. When falling the body is forced downward due to the rotationally generated force as by the rotating body, it is directly proportional to two key parameters: surface area and the rotating momentum of the rotating body. For a falling body, the uniform force, due to the rotating body, will act equally and in equal time intervals upon that body thus generating equal falling velocities. This impresses a whole force and a whole velocity proportional to the time and the parameters listed above. The spaces are described in proportional times are as the product of the velociites and the times in direct proportion to the mentioned parameters, and the squares of the times. When a body is projected upwards, this uniform force imposes forces and reduces velocities proportional to the times and parameters. The times of ascension, the absolute height, are both the product of the velocities and the times in direct proportion to the parameters. If this body where projected in a given direction, its motion is a direct form both the motion of its projection and its influence due to the afore mentioned parameters. If the body A where to be projected by some given force and would describe in a given time the line AB, and combined with its falling motion could describe in the given altitude AC, will complete the parallelogram ABCD. This same body, by that motion will be found to end its time and found in a given place D, and the curved line AED, which the body describes, will be defined as a parabola, where the right line AB will be a tangent at A. The height of the altitude AC, the area of the parallelogram ABCD, and the curvature of the curved line AED, all of which is determined to be in direct proportion to the magnitude of the above mentioned parameters.

Given the spherical form of the Earth, given as the perimeter *EFGHKI*, the surface area of the Earth is in direct proportion to lengths of the lines *EG* and *HK*. Increase the lengths of those lines, and increase the surface area as well as the area segments *EFG* and *HIK*. The rotation will effect the point positions of all key points, *E, F, G, H, K,* and I equally, as long as the Earth rotates parallel to the points F and I and equally parallel to the points *E, H* and *G, K.* Because it takes time for the Earth to shift rotation during each singular rotation periods there will be slight difference in surface tension, the strength of the Earth's surface gravity.

Mass versus weight: proportionality

Mass is the direct quanity of matter present in a given body. Its weight or gravitational mass is the direct ratio of its mass multiplied by the magnitude of the rotating frame of reference, in this case the Earth. If there were given pendulous bodies, whose center of oscillations are equally distant from the center of suspension, are in a direct ratio of the weights and the squared ratio of the times of the oscillations in a vacuum.

If a force acts on a mass, *M*, for a time dt, then by the second law of motion the velocity will be determined. The force though was acting both directly downward in summation to a force working at a slightly more perpendicular angle will result in downward motion.

Given the rectilinear motion in a central field of force, as determined by the rotational momentum of a rotating body, remains rectilinear if the angular momentum is zero and there is no constant of areas as a constraint.

Since this a rotational frame-dependent phenomenon, our equations representing gamma will be given as

$$\frac{1}{\sqrt{1-\dfrac{\omega^2}{\alpha^2}}} = \beta \qquad (1.101)$$

where ω is the rotational velocity and α is an arbitrary constant.

With the center of attraction, the rotation of the body in question, the equation of motion governing rectilinear motion along the vertical z-direction is

The equation representing a downward motion in terms of a dependence on the rotational momentum of a rotating body will be given as

$$\frac{d^2z}{dt^2} = -\frac{f}{z^2}(\gamma)$$

Let us take the integral

$$\frac{1}{2}\left(\frac{dz}{dt}(\gamma)\right)^2 = u\left(\frac{1}{z(\gamma)} + Q(\gamma)\right) = \frac{1}{2}v(\gamma)^2 \qquad (1.102)$$

Q is the constant of integration. We will have three cases

$$Q < 0, Q > 0 : Q = 0$$

The elliptic case

Here the length of $z = 2a$, the height of ascension before the velocity vanishes and starts its descent, is directly proportional to the defined parameters. The greater the rotation the greater the decrease in the length given by the above equation.

This corresponds to elliptical, hyperbolic and parabolic orbits in the non-rectilinear case. In the elliptical case we will have

$$\frac{1}{2}\left(\frac{dz}{dt}(\gamma)\right)^2 = u\left(\frac{1}{z(\gamma)} - \frac{1}{2a(\gamma)}\right) = u\left(\frac{1}{z(\gamma)} - \frac{1}{2a(\gamma)}\right) \qquad (1.103)$$

at $z = 2a$ the velocity vanishes and the motion is one of descent. Here $2a$ is produced by the gamma factor because it is a frame-dependent value. It is due to both the surface area and the magnitude of the rotation of the rotating body in question. Here the origin is at B and A is $z = 2a$ where the velocity vanishes. With the mid-point being at O of AB, we draw a semi-circle of radius $OA = OB = a(y)$. According to the figure moving in the negative direction, it will carve out the arc length $AEDB$. Given the figure below, the greater the curvature, then an increase in the length of the line CD and OD. Then following angle will increase as well; θ.

The points on the z-axis are image on points on the semicircle of the same height. The coordinates of z are

$$z(\gamma) = BC(\gamma) = a(\gamma)(1 + \cos\theta)$$
$$x(\gamma) = CD(\gamma) = a(\gamma)\sin\theta$$

(1.104)

With this we will have

$$\left(\frac{dz}{dt}\right)^2 = \frac{u(\gamma)}{a(\gamma)}\tan^2\frac{\theta}{2}$$

(1.105)

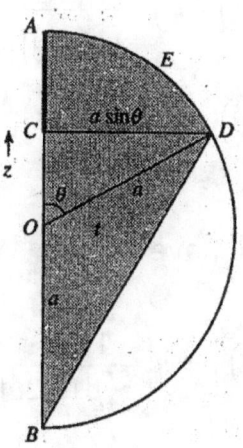

or

$$\frac{dz}{dt}(\gamma) = -\left(\frac{u(\gamma)}{a(\gamma)}\right)^{1/2}\tan\frac{\theta}{2}$$

(1.106)

the negative sign indicates a descending motion. Then we will have

$$dz(\gamma) = -a(\gamma)\sin\theta d\theta = -2a(\gamma)\sin\frac{\theta}{2}\cos\frac{\theta}{2}d\theta$$

(1.107)

equation (1.106) gives

$$dt = 2a(\beta)\left(\frac{2a(\beta)}{u(\beta)}\right)^{1/2}\cos^2\frac{\theta}{2}d\theta = a(\beta)\left(\frac{2a(\beta)}{u(\beta)}\right)^{1/2}(1+\cos\theta)d\theta \qquad (1.108)$$

integrating this equation, we will have

$$t = \left(\frac{a(\beta)}{u(\beta)}\right)^{1/2}a(\beta)(\theta+\sin\theta) \qquad (1.109)$$

where

$$t = 0$$

the body will begin its descent at $z = 2a(\gamma)$. We can rewrite the equation (1.109) in the form

$$t = \left(\frac{a(\beta)}{u(\beta)}\right)^{1/2}(arcAD+CD) = \frac{2}{\left(ua(\beta)\right)^{1/2}}(OAD+\Delta OBD)$$

$$= \frac{2}{\left(ua(\beta)\right)^{1/2}}(BDEAB) \qquad (1.110)$$

In the hyperbolic case, $Q > 0$ we will have

$$\frac{1}{2}\left(\frac{dz}{dt}(\beta)\right)^2 = u\left(\frac{1}{z(\beta)}+\frac{1}{2a(\beta)}\right) \qquad (1.111)$$

The body will have a finite velocity $(= u/a(\gamma)1/2)$ at infinity. We will now consider the case of descent starting at infinity at some point in the remote past. In the diagram, BED is the right half of the rectangular hyperbola

$$(OC)(\gamma)^2 - (CD)(\gamma)^2 = a(\gamma)^2 \qquad (1.112)$$

a is the semiaxis with the origin of z at the vertex B. The points on the $z-axis$ are imaged on the points at the same height on the rectangular hyperbola. Parametrizing the rectangular hyperbola

In the figure, the body is descending in the negative direction, then the greater the defined downward force then a proportional increase in the angle of the arc BED. The thickness defined by t will increase, the following line, $CD = a \sinh \theta$ will decrease in length, and the line CB will increase, as well.

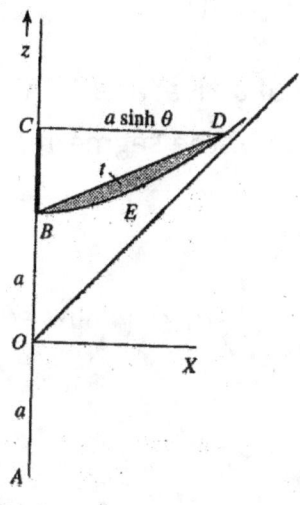

In the manner

$$OC(\gamma) = a(\gamma)\cosh\theta$$
$$CD(\gamma) = a(\gamma)\sinh\theta$$

(1.113)

we have

$$z(\beta) = BC(\beta) = a(\beta)(\cosh\theta - 1) = 2a(\beta)\sinh^2\frac{\theta}{2}$$

$$dz(\beta) = 2a(\beta)\sinh\frac{\theta}{2}\cosh\frac{\theta}{2}d\theta$$

(1.114)

and

$$\frac{dz}{dt}(\gamma) = -\left(\frac{u(\gamma)}{a(\gamma)}\right)^{1/2}\coth\frac{\theta}{2}$$

(1.115)

Here we have picked for the negative sign a descent from infinity. We than find

$$dt = -\left(\frac{u(\gamma)}{a(\gamma)}\right)^{1/2} \tanh\frac{\theta}{2}\, dz = -a(\gamma)\left(\frac{a(\gamma)}{u(\gamma)}\right)^{1/2}(\cosh-1)\,d\theta \qquad (1.116)$$

and after integration

$$t = -\left(\frac{a(\gamma)}{u(\gamma)}\right)^{1/2} a(\gamma)(\sinh\theta - \theta) \qquad (1.117)$$

where $t = 0$ is the actual time of arrival at B. This is a version of Kepler's equation for hyperbolic orbits. The area of the segment $BEDB$

is

$$\frac{1}{2}a(\gamma)^2(\sinh\theta - \theta) \qquad (1.118)$$

and

$$|t| = -t = \frac{2}{(ua(\gamma))^{1/2}}BEDB \qquad (1.119)$$

The parabolic case, $Q = 0$

In this respect the relevant equations are

$$\frac{d^2 z}{dt^2}(\gamma) = -\frac{u(\gamma)}{z^2(\gamma)}$$

$$\frac{1}{2}\left(\frac{dz}{dt}(\gamma)\right)^2 = \frac{u(\gamma)}{z(\gamma)} \qquad (1.120)$$

for a body having zero velocity at infinity.

$$dt = -\left(\frac{u(\gamma)}{a(\gamma)}\right)^{1/2} \tanh\frac{\theta}{2}\, dz = -a(\gamma)\left(\frac{a(\gamma)}{u(\gamma)}\right)^{1/2}(\cosh-1)\,d\theta$$

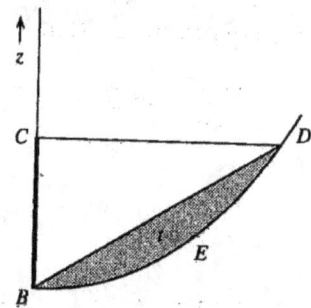

Here BED is the parabola. The degree of curvature is given as

$$x(\beta)^2 = 4a(\beta)z(\beta) \tag{1.121}$$

with its vertex at B. For a body starting its ascent from $z = 0$ at $t = 0$

$$dt = \frac{1}{\left(2u(\beta)\right)^{1/2}} z(\beta)^{1/2} \, dz$$

$$t = \frac{1}{3}\left(\frac{2}{u(\beta)}\right)^{1/2} z(\beta)^{3/2} \tag{1.122}$$

The area of the segment $BEDB$ will be

$$\frac{1}{3}a(\beta)^{1/2} z(\beta)^{3/2} \tag{1.123}$$

and

$$t = \left(\frac{2}{u(\beta)a(\beta)}\right)^{1/2} BEDB \tag{1.124}$$

The velocity of the body at C

Given the velocity of the body at C at time t from the above equations, for the elliptical case we will have

$$\left|v(C)\right| = \left(\frac{u(\gamma)}{a(\gamma)}\right)^{1/2} \tan\frac{\theta}{2} \qquad (1.125)$$

While the velocity $v(BC)$ in a circular orbit of radius BC is

$$v_{\Theta}(BC) = \frac{u(\gamma)^{1/2}}{(BC)^{1/2}} = \left(\frac{u(\gamma)}{a(\gamma)}\right)^{1/2} \frac{1}{\sqrt{(1+\cos\theta)}} = \left(\frac{u(\gamma)}{2a(\gamma)}\right)^{1/2} \sec\frac{\theta}{2} \qquad (1.126)$$

therefore

$$\frac{\left|v(C)\right|}{v_{\Theta}(BC)} = 2^{1/2}\sin\frac{\theta}{2} = \sqrt{(1-\cos\theta)} = \left(\frac{AC}{a(\gamma)}\right)^{1/2} = \left(\frac{AC}{AO(\gamma)}\right)^{1/2} \qquad (1.127)$$

Proposition *XII* . Theorem *I*

Assuming the force of surface gravity is inversely proportional to the square of the distance of the places from the centre, we seek to find the spaces which a body falling directly describes in given times as forced by the rotation of the orbiting body in question.

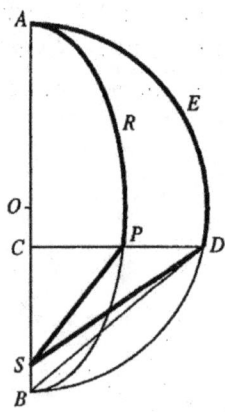

Given a body initially at rest at position A, descending directly towards the centre of attraction B, we seek to determine the times at which the body will find itself at various heights. Than given a sequence of elliptical orbits of varying eccentricities with the centre of attraction at its focus S not B, P is a point on it at some fixed height. The body at A, the aphelion point, will traverse the arc AP in a time proportional to the area of the sector $SPRA$ which is also proportional to the area of the sector $SDEA$ where D is the image of P on the auxiliary semicircle $BDEA$. This proportionality is independent of the eccentricity of the ellipse. Then when the ellipse collapses on to the right line AB and P coincides with C and S with B, we than determine that the time of descent from A to C as proportional to the area of the segment $BDEAB$. Due to the afore mentioned parameters we know that the length of the arc AB and the area of the sectors $SPRA$ and $SDEA$, along with the time of descent from A to C and the area of the segment $BDEAB$, are all directly proportional to the above mentioned parameters.

We now consider the case of a body descending from B towards infinity. Than we are given a sequence of hyperbolae of different eccentricities, with the same major axis AB. Let $BfPR$ be a hyperbola with the centre of attraction S not B and P be a point on it at some fixed height. A body starting at B, the perihelion point, will traverse the arc PfB in a time proportional to the area of the segment $BfPS$, which is also proportional to the area of the segment $BEDS$, where D is the image of P on the auxiliary rectangular hyperbola BED. This proportionality is independent of the eccentricity of the hyperbola. When the hyperbola collapses on to the z – axis and P coincides with C and S with B, we find that the time of descent from B to C is proportional to the area of the segment $BEDB$, which is proportional to the parameters.

We now consider the parabolic case. We will have a sequence of parabola with varying latera recta with vertices at the same point B, and a random but fixed parabola BED with its vertex also at B. Than if P is a point at some fixed height, on the parabola $BfPR$, then the proportionality of the distances CP and CD, where D is a point at the same height on the fixed parabola BED, the proportionality of the areas of the segments $SBfP$ and $SBED$ then follows. Then with the limit in which the parabola *BfPR* collapses on the axis, we find that the time of travel from *C* to *B* is proportional to the area of the parabolic segment *BEDB*, which is proportional to the parameters.

Proposition *XIII*. Theorem *II*

The velocity of the falling body in any place C is to the velocity of a body, describing a circle about the centre B at the distance BC, as the square root of the ratio AC, the distance of the body from the remote vertex A of the circle or rectangular hyperbola and the principle semidiameter of the figure is proportional to the parameters.

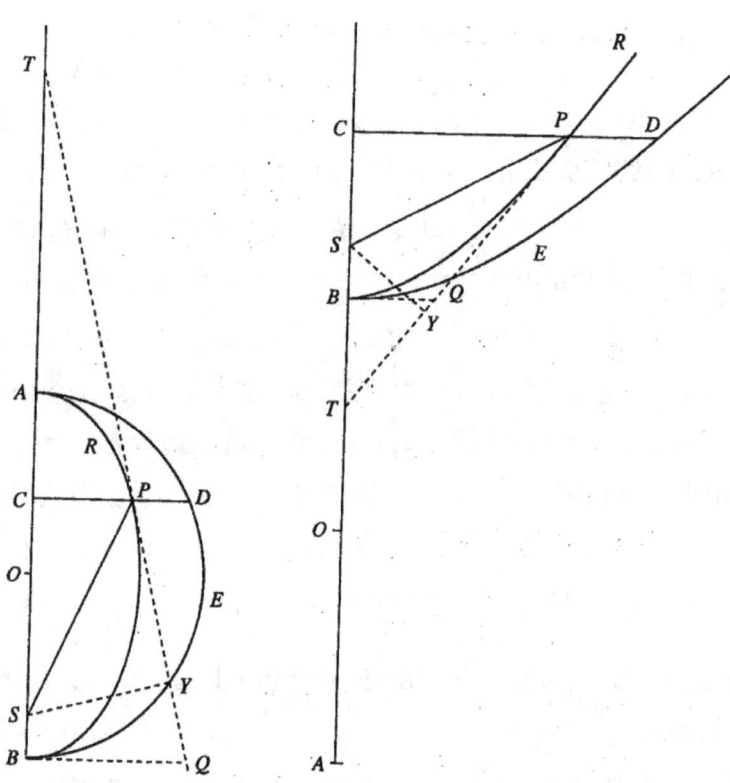

(a). The ratio of the velocity $v(P)$ in the elliptic or hyperbolic orbit to the velocity $v\ (SP)$ in a circular orbit of radius SP, is given by

$$\left[\frac{v(P)}{v\ (SP)}\right]=\frac{1}{2}L\frac{SP}{SY^2}\rightarrow\left[\frac{v(P)(\beta)}{v\ (SP)}\right] \tag{1.128}$$

given the latus rectum L and the geometry of conic sections

(b).
$$\frac{AC.CB}{CP^2}=2\frac{AO}{L}=\frac{a^2}{b^2}=\frac{2a}{2b^2/a}\rightarrow\frac{a^2(\beta)}{b^2(\beta)}=\frac{2a(\gamma)}{2b^2(\gamma)/a(\gamma)} \tag{1.129}$$

(c). We will have then

$$CO.OT\rightarrow OB^2(\beta)=OA^2(\beta)=a^2(\beta) \tag{1.130}$$

from this we obtain

$$\frac{CO}{BO}=\frac{BO}{OT}=\frac{BO+CO}{BO+OT}=\frac{BC}{BT} \tag{1.131}$$

and

$$\frac{AC}{AO}=1-\frac{OC}{AO}=1-\frac{OC}{OB}=1-\frac{BC}{BT}=\frac{BT-BC}{BT}=\frac{TC}{BQ} \tag{1.132}$$

and since $\Delta sCPT$ and BQT are similar, we have

$$\frac{AC}{AO}=\frac{TC}{BT}=\frac{CP}{BQ}:CP=\frac{BQ.AC}{AO} \tag{1.133}$$

and by combining (1.128), (1.129) and (1.133) we will have

$$\frac{1}{2}L\frac{SP}{SY^2}=\frac{SP}{SY^2}\frac{AO.CP^2}{AC.CB}=\frac{SP}{SY^2}\frac{AO}{AC.CB}\cdot\frac{BQ^2.AC^2}{AO}=\frac{BQ^2}{SY^2}\frac{SP}{BC}\frac{AC}{AO} \tag{1.134}$$

which leads to

$$\left[\frac{v(P)}{v\ (SP)}\right]^2\rightarrow\left[\frac{v(P)(\beta)}{v\ (SP)}\right]^2 \tag{1.135}$$

and passing to the limit when the ellipse or hyperbola, BPR, collapses on to the $z-$axis, P

coincides with C and S with B, we will then have

$$\left(\frac{AC}{AO}\right)^{1/2} = \frac{v(C)}{v\ (BC)} \rightarrow \frac{v(C)(\beta)}{v\ (BC)} \tag{1.136}$$

then from equation (1.136)

$$v(0) = v\ (BO) \tag{1.137}$$

Proposition *XIV* : Theorem *III*

If the figure BED is a parabola then the velocity of a falling body in a anyplace C is equal to the velocity by which a body may uniformly describe a circle about the centre B at half the interval BC.

The velocity of a body describing a parabola RPB about the centre S, in any place P, is equal to the velocity of a body uniformly describing a circle about the same centre S at half the interval SP. Given the breadth CP of the parabola by diminished in infinitum, so that the parabolic arc PfB may coincide with the right line CB, the centre S with the vertex B, and the interval SP with the BC .

Proposition *XV* . Theorem *IV*

The area of the figure DES, described by the radius SD, is equal to the area which a body with a radius equal to half the latus rectum of the figure DES, describes in the same time by uniformly revolving about the center S. All of which is directly proportional to the magnitude of the rotating body in question in relation to the motion of the falling body describing the circular motion.

C is the position of the ascending or descending body at some instant t; D is the image of C, at the same height on $SEDA$ (a semicircle, a rectangular hyperbola with semi-axis $AO = OS$, or a fixed parabola.

C and d are the positions of C and its image D at an infinitesimal time later. OKH is a circle of radius equal to the semilatus rectum $SEDA$:

K is the position at time t of a body orbiting the circle OKH with a uniform circular velocity v $(\beta)(SO = SK)$; and k is the position at a time δt later; TD is the tangent at D to $SEDA$; and SY is perpendicular to the tangent.

Case 1. We consider both the elliptical and hyperbolic cases at the same time. In both instances the semilatus rectum of $SEDA$ is $SEDA = \dfrac{1}{2} SA = SO = SK$.

1. Proof: (a). given

$$\frac{TC}{TD} = \frac{Cc}{Dd} \qquad (1.138)$$

and

$$\frac{TD}{TS} = \frac{CD}{SY} \text{ (by the similarity of } \Delta s TDC \text{ and } TSY \text{)} \qquad (1.139)$$

then multiplying these last two equations we will have

$$\frac{TC}{TS} = \frac{Cc.CD}{SY.Dd} = \frac{AC}{AO} = \frac{AC}{SK} \qquad (1.140)$$

hence:

$$\frac{AC}{SK} = \frac{Cc.CD}{SY.Dd} \qquad (1.141)$$

(b). By Proposition XIII, equation (1.136), we will have

$$\frac{AC}{AO} = \frac{AC}{SK} = \left[\frac{v(C)}{v\ (CS)} \right]^2 \rightarrow \left[\frac{v(C)(\beta)}{v\ (CS)(\beta)} \right]^2 \qquad (1.142)$$

and

$$\frac{SK}{CS} = \left[\frac{v(CS)}{v\ (SK)}\right]^2 \to \left[\frac{v(CS)(\beta)}{v\ (SK)(\beta)}\right]^2 \tag{1.143}$$

then

$$\frac{AC}{CS} = \left[\frac{v(C)}{v\ (SK)}\right]^2 \to \left[\frac{v(C)(\beta)}{v\ (SK)(\beta)}\right]^2 \tag{1.144}$$

(c). By using this relation $AC.CS = CD^2$ we will have in the geometry of conic sections

$$\frac{AC}{CD} = \frac{v(C)}{v\ (SK)} \to \frac{v(C)(\beta)}{v\ (SK)(\beta)} \tag{1.145}$$

but since

$$Cc = v(C)\delta t \to \{v(C)(\beta)\}\,\delta t$$
$$Kk = v\ (SK)\delta t \to \{v\ (SK)(\beta)\}\,\delta t \tag{1.146}$$

hence by (1.145)

$$\frac{AC}{CD} = \frac{Cc}{Kk} : Ac.Kk = CD.Cc \tag{1.147}$$

and finally

$$SK.Kk = Sy.Dd \tag{1.148}$$

equation (1.148) implies that

$$\text{area of sector } SKk = \text{area of segment } SDd$$

and both areas are described in the same time interval δt.

Therefore the magnitude of the equal areas SKk and SDk formed in the interval of time δt is diminished, and their number increased in infinity, and there being a ratio of equality, the entire areas formed are always equal.

Case 2.   We now focus on the parabolic case.

Given equation (1.141), namely

$$\frac{TC}{TS} = \frac{Cc.CD}{SY.Dd}$$

(1.149)

where the geometrical property of the parabola is

$$CS = ST = \frac{1}{2}TC$$

(1.150)

and

$$SY.Dd = \frac{1}{2}CD.Cc$$

(1.151)

then by Proposition *XIV*, we will have

$$v(C)(\beta) = v\ (\beta)\left(\frac{1}{2}SC\right)$$

(1.152)

therefore

$$\left(\frac{SK}{\left(\frac{1}{2}SC\right)}\right)^{1/2} = \left(\frac{4SK^2}{2SK.SC}\right)^{1/2} = \frac{2SK}{CD} = \frac{v\left(\frac{1}{2}SC\right)}{v\ (SK)} \rightarrow \frac{v\ (\beta)\left(\frac{1}{2}SC\right)}{v\ (\beta)(SK)}$$

(1.153)

then by the equation of the parabola

$$CD^2 = 2SK.SC = 4aSC \rightarrow 4a(\gamma)SC$$

(1.154)

then by equations (1.152) and (1.153)

$$\frac{2SK}{CD} = \frac{Cc}{Kk} = \frac{v(C)}{v\ (SK)} \rightarrow \frac{v(C)(\beta)}{v\ (SK)(\beta)}$$

(1.155)

Proposition *XVI.*   Theorem *V*

We seek to determine the times of descent of a body falling from the location A *.*

The time of descent from rest at A to C, is proportional to the area of the segment SAD, which is directly proportional to the parameters.

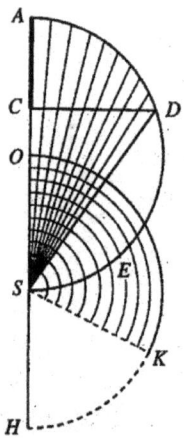

Proposition *XVII.* Theorem *VI*

To find the times of ascent or descent of a body projected upwards or downwards from a given location.

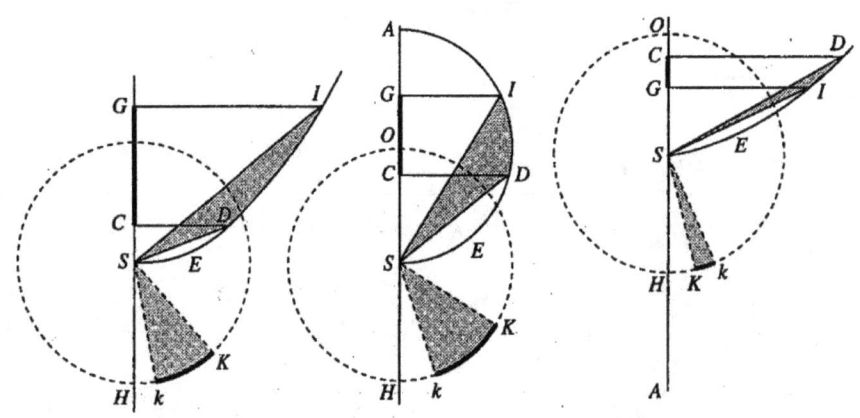

If a body is projected upwards or downwards with a given velocity we will have

$$\frac{AG}{\frac{1}{2}AS} = \frac{AS-SG}{\frac{1}{2}AS} = \left[\frac{v(G)}{v_{\Theta}(SG)}\right]^2 \rightarrow \left[\frac{v(G)(\beta)}{v_{\Theta}(SG)(\beta)}\right]^2 \qquad (1.156)$$

where $v(SG)$ is the velocity of uniform motion in a circle of radius SG, where S is the known centre of attraction. We will then have

$$2\frac{SG}{AS} = 2 - \left[\frac{v(G)}{v_{\Theta}(SG)}\right]^2 \rightarrow 2 - \left[\frac{v(G)(\beta)}{v_{\Theta}(SG)(\beta)}\right]^2 \qquad (1.157)$$

a finite positive solution exists for AS if and only it

$$\left[\frac{v(G)(\beta)}{v_{\Theta}(SG)(\beta)}\right]^2 < 2 \qquad (1.158)$$

The initial state, the velocity $v(G)$ at height SG, will belong to the elliptic class. By equation (1.157) A is infinitely remote if

$$\left[\frac{v(G)(\beta)}{v_{\Theta}(SG)(\beta)}\right]^2 = 2 \qquad (1.159)$$

this is the condition for the solution to belong to the parabolic case and the body comes to rest at infinity. This subsequent motion will belong to the hyperbolic case if,

$$\left[\frac{v(G)(\beta)}{v_{\Theta}(SG)(\beta)}\right]^2 > 2 \qquad (1.160)$$

A is the remote vertex of the rectangular hyperbola SED. The class of solutions to which the subsequent motion of a body, with an initial velocity $v(G)$ at height G, will be part to a location of A in the elliptic and hyperbolic cases determined where we find the times of ascent and descent to another specified height C, all proportional to the parameters.

Pause

Proposition *XVIII* . Theorem *VII*

Given the parameters are proportional to the altitude or distance of places from the centre of the times and velocities of falling bodies, and the spaces by which they describe, are directly proportional to the arcs and the sines and versed sines of the arcs. The bodies will accelerate with a downward force directly proportional the magnitude of the rotational force of the rotating body, the rotating frame of reference.

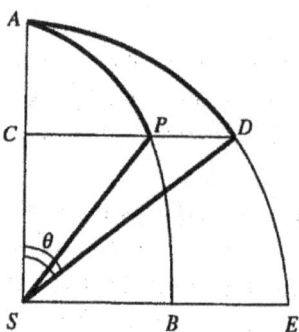

We seek to find the solution to the problem of a body, originally at rest at location A, descending directly towards the centre of attraction S. Given a sequence of ellipses of varying eccentricities and with the same semi-major axis SA. Let APB be the first quadrant of a singular ellipse and P a point on it at some fixed height. A body at A will describe the arc AP in a time proportional to the area of the sector SPA, which is also proportional to the area of the sector SDA, all of which is directly proportional to the parameters. The latter proportionality is independent of the eccentricity of the ellipse. Then passing to the limit when the ellipse collapses on to the right line AS and P coincides with C, we find that the time of descent from A to C is proportional to the area of the sector SDA, all of which is proportional to the parameters. The velocity v along the elliptical orbit is proportional to CD and which is independent of the eccentricity of the ellipse. Than the velocity acquired by the body during its descent from A to C is proportional to CD, which is proportional to the parameters.

The initial-value problem

Proposition *XIX*.   Theorem *VIII*

    Given the known parameters, and the quadratures of curvilinear figures, we seek to find the velocity of a body, ascending or descending in a right line, in the many places in which it passes, and the time in which it will arrive at any a given place.

Given the equation

$$\frac{d^2z}{dt^2} = -(z)(\beta) \tag{1.161}$$

where $f(z)$ is a positive function. The velocity is proportional to the parameters at time t and location z. The solution to this problem will be given by the integral

$$v^2 = \left[\frac{dz}{dt}\right]^2 = 2\int_z \big((f(z)\beta)\big)\,dz \tag{1.162}$$

the constant of integration is subsumed in expressing the solution as an indefinite. The solution for t is

$$t = C \pm \int^z \frac{dz}{\left[2\int_z (f(s)\beta)\,ds\right]^{1/2}} \tag{1.163}$$

If a body is subject to a force $g(z)$ different from $f(z)$, then the velocity acquired during the descent from rest at P to D is given as

$$\left[v_g(\beta)(D;P)\right]^2 = 2\int_D^P (g(z)(\beta))\,dz \tag{1.164}$$

If a body is projected from a height D with a velocity V, then the velocity at another point P is given as

$$\left[v(P)(\gamma)\right]^2 = V^2 + 2\int_P^D (f(z)\beta)\,dz \tag{1.165}$$

Proposition *XX*.   Theorem *VIV*

If a body is moved according to the parameters, and moved in any manner as such, and another body ascends or descends in a right line, and their velocities are equal in any one case of equal altitudes, their velocities will be equal at all equal altitudes.

We seek to compare the kinetic energies of two separate bodies, arriving at the same distance from the center of attraction, in this case the rotating body and the magnitude of its rotational momentum, from a common point V with the same velocity. One of them, D, is directed towards C along a radial trajectory, ADC, and the other I, along a curvilinear orbit VIK .

$$\frac{d^2r}{dt^2} = \left\{ \left(-f(r)\beta \right) + \frac{h^2}{r^3} \right\} \tag{1.166}$$

and

$$\frac{d^2z}{dt^2} = \left(f(z)\beta \right) \tag{1.167}$$

it follows then that

$$\frac{1}{2}\left(r + \frac{h^2}{r^2} \right) = \left\{ -\int^r (f(r)(\beta)) \right\} dr \tag{1.168}$$

and

$$\frac{1}{2}z^2 = -\int^z (f(z)(\beta)) dz \tag{1.169}$$

If v_r and $v_\phi = r\phi$ are the radial and transverse velocities we will have

$$r^2(\beta) + \frac{h^2}{r^2}(\beta) = v_r^2(\beta) + v_\phi^2(\beta) \tag{1.170}$$

At points I and D, the two bodies being at the same distance from the center of attraction, C, are subjected to the same frame-dependent phenomenon. They suffer equal increments in their velocity in equal times. When they arrive at E and

K,(from D and I) both equidistant from C but by infinitesimal distances, $DE = IN$ closer the increments in their velocities, will be given accelerations in the directions VDE and ITK. All of which is in the times in which they act. The body at C experiences, along the radial trajectory VDE, the change in velocity given as

$$\Delta v_{ED}\left(\beta\right)=\left(v_E-v_D\right)\beta \tag{1.171}$$

and since the component of the acceleration along TN, normal to IK, is ineffective in changing the velocity in the direction of motion, we will have

$$\Delta v_{IK}\left(\beta\right)=\left(v_K-v_I\right)\beta \tag{1.172}$$

here as the change in velocity is effected by the change in the distance from the center of rotational force, we see the separation in the contributions of the potential energy and kinetic energy to the total energy.

7

Reevaulation of the conservation of energy and the initial-value problem

Introduction

Here we focus on the continuation of the rotational frame-dependent phenomenon, deemed the surface-gravity field of force, in terms of the inverse-cube law of force.

The energy integral

Proposition *XXI* . Theorem *I*

If a given body acted upon a downward force due to the rotation of an orbiting body, is then moved in the given manner, and either descends or ascends in a right line, their velocities will be equal in the case of altitude, their velocities will also be equal at all similar altitudes, and that force is directly proportional to the magnitude of the rotational velocity of the rotating body.

If we have the given kinetic energies of two bodies, arriving at the same distance of the center of the rotating body, *C*, from a common point *V* with the same given velocity, one of them denoted *D*, toward *C*, along the radial trajectory, *ADC*, and likewise the other, denoted *I*, will be along the curvilinear orbit *VIK*.

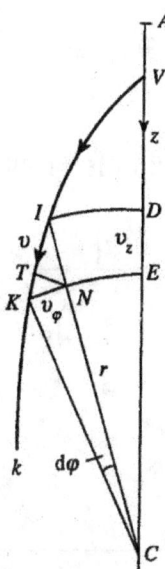

Proposition *XXII*. Theorem *II*

Given the G.R.F.F., and granting the quadratures of curvilinear figures; we seek to find the curves in which the bodies will move, as the times in the curves themselves are found.

We begin with the radial equation, given as

$$\frac{d^2r}{dt^2} = \left(\left(F_{(ij)\alpha} \oplus F_{(ij)\beta}\right)(r)\right) + \frac{[h]^2}{r^3} \tag{1.173}$$

and the angular equation

$$\frac{d\phi}{dt} = \frac{[h]}{r^2} \tag{1.174}$$

representing the law of areas. The energy integral will be given as

$$r^2 = -2\int^{r}\left(F_{(ij)\alpha} \oplus F_{(ij)\beta}\right)dr - \frac{[h]^2}{r^2} \tag{1.175}$$

we have left the lower bound of the integral unspecified. This equation can be rewritten as

$$\frac{dr}{dt} = \pm\left[-2\int^{r}\left(F_{(ij)\alpha} \oplus F_{(ij)\beta}\right)dr - \frac{[h]^2}{r^2}\right]^{1/2}$$
(1.176)

together with the equation (1.174) we will also have

$$\frac{d\phi}{dt} = \frac{[h]}{r^2\left[-2\int^{r}\left(F_{(ij)\alpha} \oplus F_{(ij)\beta}\right)dr - \frac{[h]^2}{r^2}\right]^{1/2}}$$
(1.177)

in their integrated forms we will have

$$t = \pm\int^{r}\frac{dr}{r^2\left[-2\int^{r}\left(F_{(ij)\alpha} \oplus F_{(ij)\beta}\right)dr - \frac{[h]^2}{r^2}\right]^{1/2}}$$
(1.178)

and

$$\phi = \pm\int^{r}\frac{[h]\,dr}{r^2\left[-2\int^{r}\left(F_{(ij)\alpha} \oplus F_{(ij)\beta}\right)dr - \frac{[h]^2}{r^2}\right]^{1/2}}$$
(1.179)

Let any rotational force tend towards the center C, then we seek to find the curve $VIKk$. Let there be given the circle VR, described from the center C with a radius CV; and from the same center describe any other circles ID, KE, cutting the curve in I and K, and the right line CV in D and E. We then draw the right line $CNIX$, cutting the circles KE, VR in N and X, and the right line CYK meeting the circle VR in Y. Let the points I and K be to k. Let the point A be the place from which another body is to fall, so the place D is to gain a velocity equal to the velocity of the first body in I. As in Prop. XXXIX, the short line IK, described in the least time, will be as the velocity and the right line whose square is equal to the area $ABFD$ and the triangle ICK, proportional to the time. Therefore KN will be inversely as the altitude IC; that is (if there is any quantity Q, and the altitude IC be called A) as Q/A. This quantity Q/A called Z.

Then

$$\sqrt{ABFD:Z} = IK:KN \rightarrow \frac{\sqrt{ABFD}}{h/r} = \frac{IK}{IN} = \frac{\sqrt{(v_r^2 + v_\phi^2)}}{r\phi} = \frac{v_S}{v_\phi}$$

$$\rightarrow \frac{\sqrt{(v_r^2(\beta) + v_\phi^2(\beta))}}{r\phi} = \frac{v_S(\beta)}{v_\phi(\beta)}$$

(1.180)

then

$$ABFD:Z^2 = IK^2:KN^2$$

(1.181)

by subtraction

$$ABFD - ZZ:ZZ = IN^2:KN^2$$

(1.182)

we will then have

$$\left.\begin{array}{l} ABFD:Z^2 = IK^2:KN^2 \\ ABFD - ZZ:ZZ = IN^2:KN^2 \end{array}\right] \rightarrow \frac{ABFD - h^2/r^2}{h^2/r^2} = \frac{IN^2}{KN^2} = \frac{v_r^2}{v_\phi^2} \rightarrow \frac{v_r^2(\beta)}{v_\phi^2(\beta)}$$

(1.183)

and

(1.184)

$$\left.\begin{array}{l} \sqrt{(ABFD - ZZ):Z} : \dfrac{Q}{A} = IN:KN \\ \\ A.KN = \dfrac{Q.IN}{\sqrt{(ABFD - ZZ)}} \end{array}\right] \rightarrow r^2 d\phi = \pm \frac{[h]dr}{\left[-2\int\limits^r \left(F_{(ij)\alpha} \oplus F_{(ij)\beta}\right)dr - \dfrac{[h]^2}{r^2}\right]^{1/2}}$$

(1.185)

since

$$YX.XC:A.KN = CX^2:AA \rightarrow YX.XC/(r^2 d\phi) = CX^2/r^2$$

$$YX.XC = \frac{Q.IN.CX^2}{AA\sqrt{(ABFD - ZZ)}} \rightarrow YX.XC = CX^2 d\phi$$

(1.186)

$$\rightarrow \pm \frac{CX^2 hdr}{r^2\left[-2\int\limits^r \left(F_{(ij)\alpha} \oplus F_{(ij)\beta}\right)dr - \dfrac{[h]^2}{r^2}\right]^{1/2}}$$

In the perpendicular DF, let there be taken continually Db, Dc equal to

(i):
$$\frac{1}{2}h\frac{dt}{dr} \rightarrow \frac{h}{2\left[-2\int^{r}\left(F_{(ij)\alpha} \oplus F_{(ij)\beta}\right)dr - \dfrac{[h]^{2}}{r^{2}}\right]^{1/2}}$$

(ii):
$$\frac{1}{2}CX^{2}\frac{d\phi}{dr} \rightarrow \frac{hCX^{2}}{2r^{2}\left[-2\int^{r}\left(F_{(ij)\alpha} \oplus F_{(ij)\beta}\right)dr - \dfrac{[h]^{2}}{r^{2}}\right]^{1/2}}$$

let the curved lines *ab, ac,* the loci of the points *b* and *c* be described: and from the point *V* let the perpendicular *Va* be erected to the line *AC* , cutting of the curvilinear areas *VDba,VDca* and let the ordinates *Ez, Ex* be erected as well. Because the rectangle *Db.IN* or *DbzE* is equal to half the rectangle *A.KN* or to the triangle *ICK* ; and the rectangle *Dc.IN* or *DcxE* is equal to half the rectangle *YX.XC* or to the triangle *XCY* ; that is because the bodies *DbzE* , *ICK* of the areas *VDba* , *VIC* are always equal: and the nascent bodies *DcxE* , *XCY* of the areas *VDca* , *VCX* are always equal: then the generated area *VDba* will be equal to the generated area *VIC* and thus proportional to the time; and the generated area *VDca* is equal to the generated sector *VCX* . Therefore any time be given during which the body has been moving from *V* , there will also be given the area proportional to *VDba* ; and then given the altitude of the body *CD* or *CI* ; and the area *VDca* , and the sector *VCX* equal thereto, together with its angle *VCI* . But the angle *VCI* and the altitude *CI* being given, there is likewise given the place *I* , in which the body will be found at the end at that time.

Corollary I

Given the general laws of the G.R.F.F., we will move from an inverse-square law of force to an inverse-cube law of force.

The line of apsides is determined by the integral form of the field equations.

We begin with the energy integral

$$r = \frac{2}{r} - \frac{[h]^2}{r^2} + C \tag{1.187}$$

here C is the constant of integration. This is appropriate for the inverse-square law of force.

By Corollaries I and II in Proposition XXII we will have

$$Cr^2 + 2r - [h]^2 = 0$$
$$r = 0 \tag{1.188}$$

Considering the quadratic equation above allows two real roots given as

$$r_1 = a(\beta)(1 - e\beta)$$
$$r_2 = a(\beta)(1 + e\beta) \tag{1.189}$$

then from equation (1.188) we will then have

$$[h]^2 = \frac{2r_1 r_2}{r_1 + r_2} = a(\gamma)\left(1 - \{e\}^2\right)$$
$$C = -\frac{1}{a(\gamma)} \tag{1.190}$$

then inserting these values into equation (1.187) we find

$$r^2 = 2\frac{(r_2 - r)(r - r_1)}{r^2(r_1 + r_2)} = \frac{1}{a(\gamma)r^2}\left[-r^2 + 2a(\gamma)r - a(\gamma)^2\left(1 - \{e\}^2\right)\right] \tag{1.191}$$

With the expression for r^2 the solutions for (1.178) and (1.179) take the following forms

$$\pm t = a(\gamma)^{1/2} \int \frac{r\,dr}{\left[-r^2 + 2a(\gamma)r - a(\gamma)^2\left(1 - \{e\}^2\right)\right]^{1/2}} \tag{1.192}$$

and

$$\pm \phi = [h]a(\gamma)^{1/2} \int \frac{dr}{\left[-r^2 + 2a(\gamma)r - a(\gamma)^2\left(1 - \{e\}^2\right)\right]^{1/2}} \tag{1.193}$$

then evaluating these integrals, we will find

$$t = a(\gamma)^{1/2} \left\{ a(\gamma)\left(\frac{1}{2}\pi - \sin^{-1}\frac{a(\gamma)-r}{a(\gamma)\{e\}} \right) + \left[a(\gamma)(1+\{e\})-r \right]^{1/2}\left[r - a(\gamma)(1-\{e\}) \right]^{1/2} \right\} \quad (1.194)$$

and

$$\phi = \frac{1}{2}\pi + \sin^{-1}\frac{r - a(\gamma)\left(1-\{e\}^2\right)}{r\{e\}} \quad (1.195)$$

with the initial conditions

$$t = 0 : \phi = 0 : r = a(\gamma)\left(1-\{e\}\right) \quad (1.196)$$

then the initial equation can be written as

$$\frac{a(\gamma)\left(1-\{e\}^2\right)}{r} = 1 + \{e\}\cos\phi \quad (1.197)$$

Motion under an inverse-cube law of the G.R.F.F. field

We begin with the energy integral

$$-2\int \frac{dr}{r^3} = \frac{1}{r^2} + C \rightarrow r^2 + \frac{[h]}{r^2} \quad (1.198)$$

in its integrated form we will have then

$$r^2 = \left(1-[h]^2\right)\frac{1}{r^2} + C \quad (1.199)$$

C is the constant of integration. We will have then

$$\frac{d\phi}{dt} \rightarrow \frac{[h]}{r^2} \quad (1.200)$$

With the initial conditions

$$r = 0 : t = 0 \quad (1.201)$$

for some finite

$$r = \eta > 1 \qquad (1.202)$$

(when $r = 0$ at infinity must be considered separately)

From equation (1.199) two individual cases need to be distinguished:

$$[h]^2 < 1 : [h]^2 > 1 \qquad (1.203)$$

Case. 1

for $[h]^2 < 1$ we will have for equation (1.199)

$$r^2 = \left(1 - [h]^2\right)\left(\frac{1}{r^2} - \frac{1}{\eta^2}\right) \qquad (1.204)$$

With the requirement of equation (1.202) we will have the following conditions

$$r < 0 : \phi > 0 : r < \eta \qquad (1.205)$$

then from equations (1.204) and (1.200) we will find

$$\frac{dt}{dr} = -\frac{\eta}{\sqrt{\left(1 - [h]^2\right)}} \frac{r}{\sqrt{\left((\eta)^2 - r^2\right)}} \qquad (1.206)$$

and

$$\frac{d\phi}{dr} = -\frac{[h]\eta}{\sqrt{\left(1 - [h]^2\right)}} \frac{r}{r\sqrt{\left((\eta)^2 - r^2\right)}} \qquad (1.207)$$

the solutions to these equations are

$$t = +\frac{(\eta)}{\sqrt{\left(1 - [h]^2\right)}} \sqrt{\left((\eta)^2 - r^2\right)} \qquad (1.208)$$

and

$$\phi = \frac{[h]}{\sqrt{\left(1-[h]^2\right)}} \log \frac{\eta + \sqrt{\left(\eta^2 - r^2\right)}}{r} = \frac{[h]}{\sqrt{\left(1-[h]^2\right)}} \cosh^{-1} \frac{\eta}{r} \tag{1.209}$$

we can rewrite the equation (1.208) in the following form

$$\frac{r^2}{\eta^2} + \frac{t^2}{\alpha^2} = 1 : \alpha = \frac{\eta^2}{\sqrt{\left(1-[h]^2\right)}} \tag{1.210}$$

we see that in the (t,r) plane, the orbit is an ellipse. In the following diagrams, the solution curves in the $(t,r)-$, $(\phi,r)-$ and the $(r,\phi)-$ planes are illustrated. Then when $[h]^2 < 1$, the body in question descends to the centre in a spiral curve.

Letting

$$\tan\theta = t/r \tag{1.211}$$

we have

$$\sec^2\theta \frac{d\theta}{dt} = \frac{1}{r} - \frac{t}{r^2}\frac{dr}{dt} \tag{1.212}$$

then by equation (1.210)

$$\frac{dr}{dt} = -\frac{\eta^2}{\frac{\eta^2}{\sqrt{\left(1-[h]^2\right)}}}\frac{t}{r} \tag{1.213}$$

then eliminating dr/dt from equation (1.212)

$$\frac{\eta}{r}\left(r^2 + t^2\right)d\theta = \frac{\eta^3}{[h]}d\phi \tag{1.214}$$

If

$$dA = \frac{1}{2}\left(r^2 + t^2\right)d\theta \tag{1.215}$$

which denotes the element of area in the $(t,r)-$plane, then

$$\frac{\eta}{r}dA = \frac{1}{2}\frac{\eta^3}{[h]}d\phi \qquad (1.216)$$

or in its integrated form

$$\int_\eta^r \frac{\eta}{r}dA = \frac{1}{2}\frac{\eta^3}{[h]}\phi : (\phi = 0, r = \eta) \qquad (1.217)$$

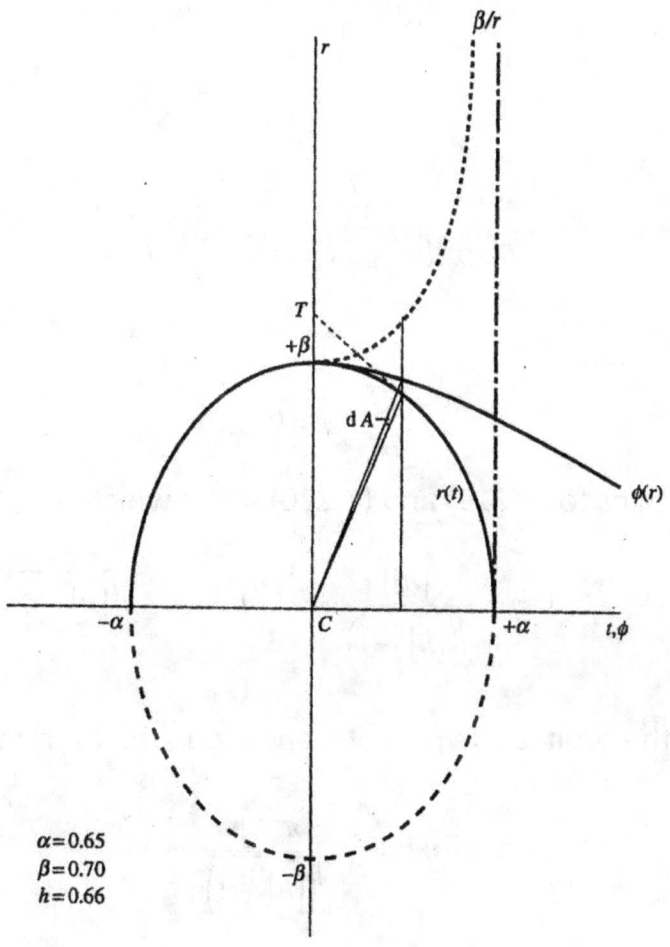

$\alpha = 0.65$
$\beta = 0.70$
$h = 0.66$

Alternatively, we will also have

$$A = \frac{1}{2}\frac{\eta^2}{[h]}\int_0^\phi rd\phi = \frac{1}{2}\frac{\eta^2}{[h]} \times \text{arc length along the orbit of the body} \qquad (1.218)$$

Equations (1.217) and (1.218) are equivalent to two of Newton's statements in this corollary. They are referred to as Newton's identities.

Case. 2 $(\beta)^2 > 1$

Here equations (1.206) and (1.207) are replaced by

$$\frac{dt}{dr} = +\frac{\eta}{\sqrt{([h]^2 - 1)}}\frac{1}{\sqrt{((r)^2 - \eta^2)}} \tag{1.219}$$

and

$$\frac{d\phi}{dr} = +\frac{[h]\eta}{\sqrt{([h]^2 - 1)}}\frac{1}{\sqrt{((r)^2 - \eta^2)}} \tag{1.220}$$

where we have required that

$$r \geq 0 : \phi \geq 0 \tag{1.221}$$

The necessary solutions to (1.219) and (1.220) along which $\phi = 0$ and $r = \eta$ are

$$\frac{r^2}{\eta^2} - \frac{t^2}{\alpha^2} = 1 : \phi = \frac{[h]}{\sqrt{([h]^2 - 1)}}\cos^{-1}\frac{\eta}{r}\left[\alpha = \eta / \sqrt{([h]^2 - 1)}\right] \tag{1.222}$$

In the (t,r)–plane the orbit is a hyperbola; and along the asymptote $r / t = \eta / \alpha$

$$\phi \to \phi_0 = \frac{1}{2}\frac{\pi[h]}{\sqrt{([h]^2 - 1)}} \tag{1.223}$$

the curve ascends indefinitely while ϕ tends to a finite limiting value.

The newly developed identities (1.217) and (1.218) are readily verifiable. In the accompanying diagrams the solution curves in the $(t,r)-, (\phi,r)-$ in the $(r,\phi)-$ plane are illustrated.

We now consider the case in which the body comes to rest at infinity.

155

Case 3.

With $r = 0$ at $r = \infty$. In this case the energy integral gives

$$r^2 = \frac{1-[h]^2}{r^2} : ([h]^2 > 1 \text{ is not allowed})$$

(1.224)

We then have the equation

$$\phi = [h]/r^2$$

(1.225)

the solutions of these equations then gives

$$r = \exp\left[\frac{\sqrt{\left(1-[h]^2\right)}}{[h]}(\phi - \phi_0)\right] = \left[2(t-t_0)\sqrt{\left(1-[h]^2\right)}\right]^{1/2}$$

(1.226)

Proposition *XXIII*.   Theorem *III*

The law of the rotational-dependent force being given, we seek to find the motion of a body setting out from a given place with a given velocity, in the direction of a right line.

This is the formulation of the initial-value problem. At a given instant of time, a body is projected from a point P, at a distance r_0 from the center S of a known rotational-dependent force, with a prescribed velocity V in some specified direction; we are to find the motion that will follow. The solution to this problem requires the lower bound of the integral equation (1.175)

$$r^2 = -2\int^r\left(\left(F_{(ij)_\alpha} \oplus F_{(ij)_\beta}\right)r\right)dr - \frac{[h]^2}{r^2}$$

(1.227)

consistently with the stated initial conditions. These are satisfied by writing

$$\frac{1}{2}\left(r^2 + \frac{h^2}{r^2}\right) = -\int_{r_0}^r\left(F_{(ij)_\alpha} \oplus F_{(ij)_\beta}\right)dr + \frac{1}{2}\left(V_r^2 + V_\phi^2\right)$$

(1.228)

where V_r and V_ϕ are the radial and transverse components of V. But since the constant of areas $[h]$ during the subsequent motion of the body must retain the value that it had at he beginning,

$$[h] = (\omega_0 \times m) \tag{1.229}$$

with this value for $[h]$, equation (1.228) gives

$$r^2 = -2\int^r \left(F_{(ij)_\alpha} \oplus F_{(ij)_\beta} \right) dr - \frac{r_0^2}{r^2} V_\phi^2 + V^2 \tag{1.230}$$

the solution for the initial-value problem follows from inserting this expression (1.227) in place of equations (1.178) and (1.179). Thus

$$t = \pm \int^r \frac{dr}{\left[-2\int_{r_0}^r \left(F_{(ij)_\alpha} \oplus F_{(ij)_\beta} \right) dr + V^2 - r_0^2 V_\phi^r / r^2 \right]^{1/2}} \tag{1.231}$$

and

$$\phi = \pm \int^r \frac{dr}{r^2 \left[-2\int_{r_0}^r \left(F_{(ij)_\alpha} \oplus F_{(ij)_\beta} \right) dr + V^2 - r_0^2 V_\phi^r / r^2 \right]^{1/2}} \tag{1.232}$$

8

On revolving orbits in terms of the G.R.F.F. potential

Introduction

Here we consider the the theorem of revolvoing orbits in terms of the G.R.F.F. potential.

Theorem of revolving orbits

Proposition *XXIV*.   Theorem *I*

It is required to make a singular body move in a given curve that revolves about the center of force, the G.R.F.F. *in the same manner as another body in the curve at rest.*

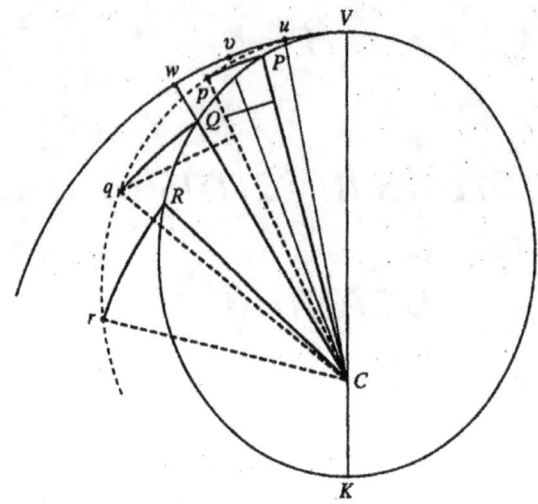

Let $VPQR...K$ be the fixed orbit described by a body under the influence of the *G.R.F.F.* centered at C and $P, Q, R, ect.$, the positions occupied by a body at a chosen instant of time. The 'revolving orbit' is obtained by imaging the points $P, Q, R, ect.$ on to the fixed orbit at the same chosen instant of time by the transformation

$$Cp = CP : \angle VCp = \alpha \angle VCP$$
$$Cq = CQ : \angle VCq = \alpha \angle VCQ, ect. \tag{1.233}$$

where α is a chosen constant. By this transformation

$$\text{area of sector } VCp = \frac{1}{2}(Cp)^2 \times \angle VCP = \frac{1}{2}\alpha(CP)^2 \times \angle VCP = \alpha \times \tag{1.234}$$

area of sector VCP

and since both sectors are described in the same time

$$\frac{VCp}{t} = \alpha \frac{VCP}{1} = \frac{1}{2}\alpha[h] \tag{1.235}$$

where

$$VCp = \text{area of sector} \tag{1.236}$$

and

$$VCP = \text{area of sector} \qquad (1.237)$$

then

$$t = \text{time in which the area is swept} \qquad (1.238)$$

where $[h]$ is the constant of areas along the fixed orbit. And since relation (1.237) is for all the chosen sectors, it follows that along the revolving orbit, as imaged, equal areas are swept in equal times; and the constant of areas is $\alpha[h]$. Therefore by the proposition (every body that moves in a curved line describred in a plane, and by a radius drawn to a point, describes about that point areas proportional to the times, determined by the *G.R.F.F.*), the revolving orbit is described under the action of the *G.R.F.F.*.

We obtain the following relations based on the transformations described:

$$\angle VCp = \alpha\angle VCq = \alpha\angle VCQ :$$
$$\angle PCp = \angle VCu; \angle QCq = \angle VCc :$$
$$\left\{ \begin{array}{l} \angle PCp = \angle VCp - \angle VCP = (\alpha-1)\angle VCP \\ \angle QCq = \angle VCq - \angle VCQ = (\alpha-1)\angle VCQ \end{array} \right\} \qquad (1.239)$$

and

$$\angle PCQ = \angle VCQ - \angle VCP = \frac{1}{\alpha}[\angle VCq - \angle VCp] = \frac{1}{\alpha}\angle pCq \qquad (1.240)$$

or

$$\angle pCq = \alpha\angle PCQ \qquad (1.241)$$

In the fixed orbit VPK, let the body P revolve moving from V towards K. From the center C let there be continually drawn Cp equal to CP, making the angle VCp proportional to the angle VCP; and the area which the line Cp describes will be to the area VCP, which the line CP describes at the same time, as the velocity of the describing Cp to the velocity of the describing line CP; that is, as the angle VCp to the angle VCP, then in a given ratio, and then proportional to the time. The area described by the line Cp in a fixed plane is proportional to the time, then a body being acted on by the G.R.F.F., may revolve with the point p in the curved line which the same point p, by the same method just defined, may be made to describe in a fixed plane. The angle VCu is equal to the angle PCp, and the line Cu equal to CV, and the figure uCp equal to the figure VCP, and the body always in the point p, will move in the perimeter of the revolving figure uCp, and will describe its revolving arc up in the same time that the other body P describes the similar and equal arc VP in the fixed figure VPK.

Proposition *XXV*.   Theorem *II*

The difference of the forces, the G.R.F.F., *by which two bodies are to moved equally, on in a fixed orbit, the other in the same orbit revolving, varies as the cube of their common altitudes.*

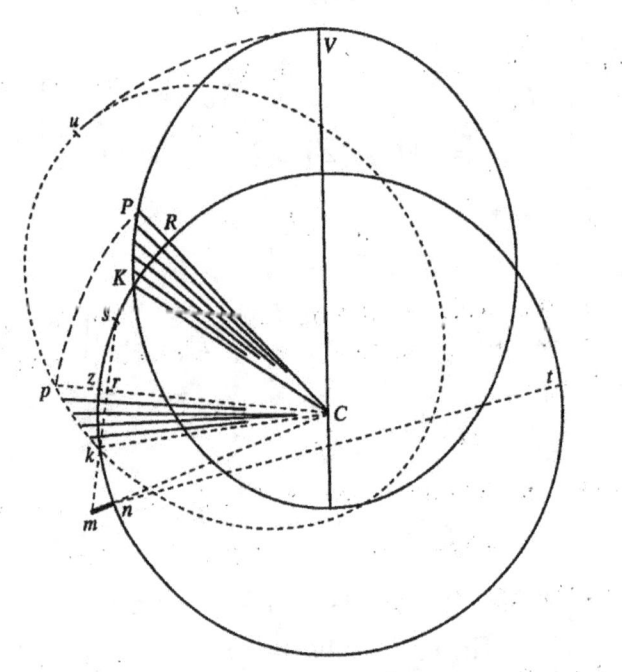

This diagram, similar to the one in Proposition *XXIV*, is in the same context. VPK is the fixed orbit; the vertex V and the position P being imaged on u and p in the congruent revolving orbit upk. Then by equations (1.233) and (1.241) of the prior section

161

$$Cu = CV : Cp = CP : \angle VCu = \angle PCp \tag{1.242}$$

Let K be a neighboring point, infinitely close to P; and if n is its image

$$Cn = CK \tag{1.243}$$

and by equation (1.241)

$$\angle pCn = \alpha \angle PCK \tag{1.244}$$

let k be the intersection of the circle of radius $CK = Cn$ with the orbit passing through u and p congruent to VPK. By this the sectors CPK and Cpk are congruent.

$$CPK = Cpk \tag{1.245}$$

draw kr perpendicular to Cp and extend it to intersect the circle $CKkn$ at s; then

$$kr = rs \tag{1.246}$$

On the fixed orbit, P arrives at K (after an time interval Δt). In the direction normal to the orbit at P, it will have moved a distance equal to the perpendicular distance of K from PC, that is equal to kr (to the first order) by the congruence of the figures CPK and Cpk. The imaged point, p on the 'revolving orbit' will arrive at m, the continuation of kr, where by equation (1.244)

$$rm = \alpha rk \tag{1.247}$$

mn is to continue to intersect the circle Kkn at t. By the area theorem we will have

$$\Delta pCk = \Delta PCK = \frac{1}{2} rk \cdot PC = \frac{1}{2}[h]\Delta t \tag{1.248}$$

here $[h]$ is the constant of areas along the fixed orbit; then by equation (1.247)

$$\Delta pCm = \frac{1}{2} rm \cdot PC = \frac{1}{2} \alpha rk \cdot PC = \frac{1}{2} \alpha [h] \Delta t \tag{1.249}$$

By simple geometry we will then have

$$ms = rm + rs = rm + rk = (\alpha + 1)rk = \frac{[h]\left(\frac{[h']}{[h]} + 1\right)}{PC} \Delta t$$

$$(1.250)$$

$$mk = rm - rs = rm + rk = (\alpha - 1)rk = \frac{[h]\left(\frac{[h']}{[h]} - 1\right)}{PC} \Delta t$$

The *G.R.F.F.* for both coordinates P and p will be

$$G.R.F.F(P) = \sum_{i=1}^{\theta} \left(-\cos\theta CK + \sin\theta CK\right) \cdot AK \cdot \sum_{j=1}^{\theta} \left(-\cos\theta CP + \sin\theta CKP\right)_j$$

$$\sum_{i=1}^{\theta} \left(\cos\theta CK + (-)\sin\theta CK\right) \cdot AK \cdot \sum_{j=1}^{\theta} \left(\cos\theta CP + (-)\sin\theta CKP\right)_j$$

$$(1.251)$$

$$G.R.F.F(p) = \sum_{i=1}^{\theta} \left(-\cos\theta Ck + \sin\theta Ck\right) \cdot Ak \cdot \sum_{j=1}^{\theta} \left(-\cos\theta Cp + \sin\theta Cp\right)_j$$

$$\sum_{i=1}^{\theta} \left(\cos\theta Ck + (-)\sin\theta Ck\right) \cdot Ak \cdot \sum_{j=1}^{\theta} \left(\cos\theta CP + (-)\sin\theta Cp\right)_j$$

Cor. The difference of forces, with which the body *P* revolves in a fixed orbit and the body p in a movable orbit, will be to the *G.R.F.F* , with which another body by a radius drawn to the center can uniformly describe that sector in the same time as the area *VPC* is described as $[h'][h'] - [h][h]$ to $[h][h]$.

Then

(*G.R.F.F.* at p in revolving orbit − *G.R.F.F.* at *P* in fixed orbit) $\times (\Delta t)^2 /$ (*G.R.F.F.* in circular orbit or radius $PC = pc$) $\times (\Delta t)^2 \rightarrow$ (*G.R.F.F.* at p in revolving orbit − *G.R.F.F.*

at *P* in fixed orbit) / (*G.R.F.F.* in circular orbit or radius $PC = pc$) $= \dfrac{[h']^2 - [h]^2}{[h]^2}$

where $[h']$ and $[h]$ denote the constant of areas in the revolving and in the fixed orbits.

Corollary I

We seek an expression for the G.R.F.F. *acting on a given body describing the revolving orbit when the fixed orbit is an ellipse.*

We start with

G.R.F.F. at a distance $pC = PC = A$ in the revolving orbit $= G.R.F.F.$ at a distance $PC = pC = A$ along the fixed elliptic orbit $+ \dfrac{[h']^2 - [h]^2}{A^3}$

where

$$[h]\alpha = [h'] \qquad\qquad (1.252)$$

Then by proposition XI, G.R.F.F. at P along the fixed elliptic orbit $= [h]^2 / lA^2$

since in the present notation $l = $ semilatus rectum. Hence G.R.F.F. at a distance A from the center of attraction along the revolving orbit $= \dfrac{1}{l}\left[\dfrac{[h]^2}{A^2} + \dfrac{l[h']^2 - [h]^2}{A^3} \right]$

Generally speaking, a force acts at a point P in the direction towards and away from the normal ρ, the radius of curvature that pierces the moving body through the center of line of the greater body in question. Applying this result at V

G.R.F.F. at V along a circular orbit of radius CV / G.R.F.F. at V along the fixed elliptic orbit $=$ radius of curvature of an ellipse at the vertex V / radius of curvature of circular orbit of radius CV

then for equal velocities along both orbits, equation $\dfrac{[h']^2 - [h]^2}{[h]^2}$ now gives

G.R.F.F. at V along the revolving orbit

$=$ G.R.F.F. at V along the fixed elliptic orbit

$$\times\left[1+\frac{l}{CV}\frac{[h']^2-[h]^2}{[h]^2}\right]=\frac{[h]^2}{l(CV)^2}\left[1+\frac{l}{CV}\frac{[h']^2-[h]^2}{[h]^2}\right]=\frac{1}{l}\left[\frac{[h]^2}{(CV)^2}+\frac{l}{(CV)^3}\left([h']^2-[h]^2\right)\right]$$

Corollary II

In this corollary, the explicit form of the equation of the G.R.F.F is sought when the fixed orbit is an ellipse described about its center.

We begin with

G.R.F.F. acting at P in fixed orbit $=\dfrac{[h]^2}{a^2(\gamma)b^2/(\gamma)}PC$

where the semilatus rectum is given as

$$l=\frac{b^2/\gamma}{a(\gamma)} \tag{1.253}$$

then equation $\dfrac{[h]^2}{a^2(\gamma)b^2/(\gamma)}PC$ will be given as

$$[h^2]\left(\frac{1}{l}\frac{b^2/\gamma}{a(\gamma)}\right)\frac{PC}{a(\gamma)^2\,b^2/\gamma}-\frac{[h]PC}{la(\gamma)^3} \tag{1.254}$$

then

G.R.F.F. acting at the same distance in a revolving orbit

$$=\frac{[h]^2PC}{la(\gamma)^3}+\frac{[h']^2-[h]^2}{PC^3}=\frac{1}{l}\left[\frac{[h]^2PC}{a(\gamma)^3}+l\frac{[h']^2-[h]^2}{PC^3}\right]$$

Corollary III

By equation (1.251) for G.R.F.F. we will have

$$G.R.F.F. \text{ at } P \text{ in fixed orbit} = \frac{1}{\rho \sin^3 \in} \frac{[h]^2}{PC^2} \qquad (1.255)$$

then

$G.R.F.F.$ at the same distance from C in the revolving orbit

$$= \frac{1}{\rho \sin^3 \in} \left[\frac{[h]^2}{PC^2} + (\rho \sin^3 \in) \frac{[h']^2 - [h]^2}{PC^3} \right] \qquad (1.256)$$

where ρ is the radius of curvature at P and $\in$ is the inclination of PC to the direction of motion at P. Then since along an elliptic orbit $(\rho \sin^3 \in) = l$

Cor. V. The motion of a body in a fixed orbit being given, its angular motion around a center of force may be diminished or increased in a given ratio; and then the new fixed orbits may be found in which bodies may revolve with new $G.R.F.F.$

.

Corollary IV

Let VX be perpendicular to CV; and consider a point P moving along VX with a uniform velocity. Since P moves by inertia alone, it is not subject to any other force acting on it. Hence the image point p in the revolving orbit (at a distance $Cp = CP$) will describe this orbit under the $G.R.F.F.$ inversely proportional to $(pC)^3$.

$$G.R.F.F. \text{ acting on } p \text{ at a distance } pC = [h]^2 \left(\frac{[h']^2}{[h]^2} - 1 \right) (pC)^3 \qquad (1.257)$$

where the constant of areas is given like before for $[h]$. Then

$$G.R.F.F. \text{ acting on } p \text{ at a distance } pC = [h]^2 \left(\frac{[h']^2}{[h]^2} - 1 \right) / r^3 \qquad (1.258)$$

Proposition *XXVI*. Theorem *III*

To find the motion of apsides in orbits approaching very near to circles.

We consider her orbits very near to circles described under arbitrary $G.R.F.F.$ to nearly circular elliptical orbits described under inverse-square forces; and we seek to determine the precession of the line apsides under these circumstances.

By Proposition *XXV* we begin with

$$G(r)' = G(r) + \frac{[h']^2 - [h]}{r^3} \tag{1.259}$$

For an elliptical orbit described under an inverse-square law of both equal attraction and repulsion

$$G(r) = \frac{[h]^2}{lr^2} \tag{1.260}$$

where l is the semilatus rectum of an ellipse of eccentricity e and semiaxes a and b, is

$$l = \frac{b^2/\gamma}{a(\gamma)} = a(\gamma)\left(1 - \{e\}^2\right) \tag{1.261}$$

then equation (1.259) will be given as

$$G'(r) = \frac{1}{l}\left[\frac{[h]}{r^2} + \frac{l\left([h']^2 - [h]^2\right)}{r^3}\right] = \frac{1}{l}\left[\frac{l\left([h']^2 - [h]^2\right) + r[h]^2}{r^3}\right] \tag{1.262}$$

for a very nearly circular orbit we may write

$$r = r_{max} - X = T - X$$
$$T = a(1+e) \tag{1.263}$$

We then will have

$$lG(r) = C(r)r^{-3} = C(T-X)r^{-3} = \left[C(T)-XC'(T)\right]r^{-3}$$
$$\rightarrow \frac{1}{r^3}\left[l[h']^2 - [h]^2 + (T-X)[h]^2\right]$$

(1.264)

and since we are dealing with nearly circular orbits we can expand $C(T-X)$ in a Taylor series. We retain only the zero and the first-order terms.

This gives us

$$lG(r) = \left[C(T)-XC'(T)\right]r^{-3}$$

(1.265)

which gives us for equation (1.264)

$$l\left([h']^2 - [h]^2 + (T-X)[h]^2\right) = C(T) - XC'(T)$$

(1.266)

then we will have

$$\frac{1}{C(T)}\left[l\left([h']^2 - [h]^2\right) + T[h]^2\right] - \frac{X[h]^2}{C(T)} = 1 - \frac{XC'(T)}{C(T)}$$

(1.267)

The first terms of both sides of the equation are of zero order while the second terms are of the first order. Then with

$$\frac{1}{C(T)}\left[l\left([h']^2 - [h]^2\right) + T[h]^2\right] = 1$$

(1.268)

Since l and T to zero-order are both equal to the semimajor axis a, we find

$$\frac{T[h']^2}{C(T)} = 1$$

(1.269)

then since

$$[h']^2 = C(T) : [h] = C'(T)$$

(1.270)

equation (1.267) will read as

$$\frac{1}{[h']^2}\left[l\left([h']^2 - [h]^2\right) + T[h]^2\right] - \frac{X[h]^2}{[h']^2} = 1 - \frac{X[h]^2}{[h']^2}$$

(1.271)

and we will have for

$$lG(r) = \left[[h']^2 - X[h]^2\right]r^{-3} = \frac{1}{r^3}\left[l[h']^2 - [h]^2 + (T-X)[h]^2\right] \qquad (1.272)$$

then for $T = 1$

$$\alpha^2 = \frac{[h']^2}{[h]^2} \qquad (1.273)$$

Example 1

Here a revolving orbit under the a uniform $G.R.F.F.$ leads to

$$[h']^2 = \left\{a(1 + \{e\})\right\}^3 \qquad (1.274)$$

Since a body, in a fixed ellipse, in descending from the upper to the lower apse, describes some form of an angle, say 180 degrees, the other body that moves in a movable ellipse, and in a fixed plane will in its descent from the upper to lower apse describe a given angle $\angle VCp$. Therefore a body revolving with a uniform $G.R.F.F.$ in an orbit nearly circular, will always describe an angle given as $\angle VCp$ at the centre, moving from the upper apse to the lower apse when it has once described that afore mentioned angle. It returns to the upper apse when it has described that aforementioned angle, and so on.

Example 2

The $G.R.F.F.$ is proportional to r'^{n-3} which leads to

$$[h']^2 = \left\{a(1 + \{e\})\right\}^n \qquad (1.275)$$

Therefore since the angle VCP, described in the descent of the body from the upper apse to the lower apse in an ellipse, is of 180 degrees, the angle VCp, described in the descent of the body from the upper apse to the lower apse in an orbit nearly circular which the body describes with a *G.R.F.F.* proportional to the power A^{n-3}, will be equal to an angle of $180'/\sqrt{n}$ and this angle being repeated, the body will return from the lower to the upper apse and so on.

Example 3

For the chosen $C(T)$ we will have

$$\alpha^2 = \frac{[h']^2}{[h]^2} = \left[\frac{b\{a(1+\{e\})\}^m \pm c\{a(1+\{e\})\}^n}{mb\{a(1+\{e\})\}^m \pm nc\{a(1+\{e\})\}^n} \right]_{T=1} = \left[\frac{bT^m \pm cT^n}{mbT^m \pm ncT^n} \right]_{T=1} = \frac{b\pm c}{mb\pm nc} \quad (1.276)$$

therefore since the angle *VCP* between the upper and the lower apse, in a fixed ellipse is 180, the angle *VCp* between the same apsides in an orbit which a body describes with a *G.R.F.F.*, that is as $\left(bA^m \pm cA^n\right)/A^3 = 180/(b\pm c)(mb\pm nc)$

Cor. *I* Hence if the *G.R.F.F.* be as any power of altitude, that power may be derived from the motion of the apsides.

If a body describing the revolving orbit returns to the same apse m times when the fixed orbit completes n complete revolutions, then

$$\angle VCp = \frac{m}{n} \angle VCP : \alpha = \frac{m}{n} = \frac{[h']}{[h]} \quad (1.277)$$

we then confer from equations from example 2

$$C(A) = A^{n^2/m^2} = A^{[h]^2/[h']^2} \quad (1.278)$$

A body revolving with the *G.R.F.F.*, and parting from the apse, it then begins to ascend, will never arrive at the lower apse or least altitude, but will then descend to the center, describing a curved line treated in Cor. *III*, Prop. *XLI*. But if it were to part slightly from the lower apse, begin to ascend a little, it will ascend into infinity. It will then never come to the upper apse; will describe the curved line spoken of in Cor. *VI*, Prop. *XLIV*. If the prior mentioned force in its recess were to from the center to decrease in a greater than a cubed ratio of the altitude, the body at its separating from the apse, will either descend to the center or ascend into infinity, according as it descends or ascends at the start of its motion. But if the prior mentioned force in its recess from the center either decreases in a less than a cubed ratio of the altitude, or increases in any ratio of the altitude whatsoever, the body will never descend to the center, but will arrive at some at the lower apse; and if the body alternatively ascending or descending from one apse to another never comes to the center, then either the force increases in the recess from the center, or it decreases in a less than a cubed ratio of the altitude; and the sooner the body returns from one apse to another, the farther is the ratio of the forces from the cubed ratio.

9

Momentary Break

Introduction

Here is the continuation of the G.R.F.F. in terms of ellipses.

Proposition *XXVII*.   Theorem *I*

Given the G.R.F.F.*, and the center of force, and any plane being given in which the body revolves, and the quadratures of curvilinear figures allowed; we seek to determine the motion going off from some given place with a given velocity, in the direction of a given right line in that plane.*

Here the center of force S is of the plane (X,Y) in which the motion of the body at P takes place.

Draw *SC* perpendicular to the plane (X,Y). The G.R.F.F. at a point Q is

$$\left(F_{(ij)_\alpha} \oplus F_{(ij)\beta}\right)v\cdot dr\frac{r}{r} \tag{1.279}$$

acts in the direction towards and away directly along QC and depends only on the distance QC. We resolve this force along QC as the center line of force in the plane (X,Y) and in the direction both away and towards from QC normal to the plane (X,Y). The latter force of magnitude

$$\left(\left(F_{(ij)_\alpha} \oplus F_{(ij)\beta}\right)v\cdot dr\right)\sin\phi \tag{1.280}$$

will not affect the motion confined to the plane (X,Y); just the force in the direction both towards and away from CQ will affect its motion. The force in terms of this will be given as

$$\left(\left(F_{(ij)_\alpha} \oplus F_{(ij)\beta}\right)v \cdot dr\right)\cos\phi\frac{\varpi}{\omega} = \left(\left(F_{(ij)_\alpha} \oplus F_{(ij)\beta}\right)\sqrt{\left(CQ^2+\varpi^2\right)}\right)\frac{\varpi}{\sqrt{\left(CQ^2+\varpi^2\right)}}\frac{\varpi}{\omega} \quad (1.281)$$

here ϖ denotes the radial distance, QC in the plane (X,Y) and $\frac{\varpi}{\omega}$ is the unit vector in the radial direction in the same plane. Since QC is constant, the force will be given

$$\left(\left(F_{(ij)_\alpha} \oplus F_{(ij)\beta}\right)\sqrt{\left(CQ^2+\varpi^2\right)}\right)\frac{\varpi}{\sqrt{\left(CQ^2+\varpi^2\right)}} \quad (1.282)$$

is a function of the radial distance only in the plane (X,Y).

The action of the other force in terms of QC, coinciding with the position of the plane itself, both attracts and repels simultaneously the body from the point C in that plane, therefore causes the body to move in the plane in the same manner as if the force direct along the line SQ did not exist, then the body were to revolve in free space about the center C along the line QC.

Proposition XXVIII. Theorem II

Given that the G.R.F.F. *to be proportional to the distance of the body from the center; all bodies revolving in any planes will describe ellipses, and complete their revolutions in equal times; and those which move in right lines, will complete their several periods of going and returning in the same times.*

Therefore the forces with which the bodies found in the plane (X,Y) are attracted and repelled towards the center point C, are in proportion to the distances equal to the forces with which the same bodies are both attracted and repelled from the center S; and therefore the bodies will move in the same times and in the same figures, in any plane (X,Y) about the point C, as they would in free spaces about the center S; and therefore they will in equal times either describe ellipses in that plane about the center C, or move to and fro in right lines passing through the center C in that plane, completing the same periods of time in all cases.

Proposition *XXIX*.   Theorem *III*

Given the G.R.F.F., *and the centre of force, and any plane in which the body revolves, and the quadratures of curvilinear figure allowed we seek to determine the motion of a body going off from a given place with a given velocity in the direction of a right line in that plane.*

10

The two-body problem reevaluated

Introduction

Here we focus on the gravitational behavior between bodies in regards to an immovable center. Given the behaviors between multiple bodies are always reciprocal and equal, then by Law III, there are two bodies, neither the larger mass and the smaller revolving mass are at rest, but both will revolve around a common center of gravity. If there are more bodies, moving in relation to a larger center of mass, and are also likewise affected by each other, then will be so moved among themselves that the common center of gravity will either be at rest or moved in a straight line. Here we will focus on the nature of the G.R.F.F.

General theorems

Proposition *XXX*. Theorem *I*

Two bodies of varying mass will attract each other mutually and describe varying figures about their common center of gravity according to the G.R.F.F.. The smaller the difference in mass between the two bodies than the smaller the difference in orbiting revolutions between them and the larger the vibratory revolution of the larger mass.

Let $M1$ and $M2$ denote two inertial masses of varying mass. The attractive force between them will result according to the original G.R.F.F. This will be given as

$$F(M_1, M_2; r_{12})(F > 0) \tag{1.283}$$

F is not symmetrical in $M1$ and $M2$ and r_{12} is the distance between them. The equations governing the motions of both bodies will given as

$$M_1 \frac{d^2 r_1}{dt^2} = -F(M_1, M_2; r_{12}) \frac{r_1 - r_2}{r_{12}}$$

$$M_2 \frac{d^2 r_2}{dt^2} = -F(M_1, M_2; r_{12}) \frac{r_1 - r_2}{r_{12}}$$

(1.284)

The uniform motion in a straight line of the centre of gravity will be given as

$$C = \frac{M_1}{M_1 + M_2} r_1 + \frac{M_2}{M_1 + M_2} r_2$$

(1.285)

then

$$M_1 \frac{d^2 r_1}{dt^2} + M_1 \frac{d^2 r_2}{dt^2} = 0$$

(1.286)

and

$$(M_1 + M_2) C = M_1 r_1 + M_2 r_2 = (M_1 + M_2) Vt + A$$

(1.287)

The motion of the two bodies around one another are given by the equation

$$\frac{d^2}{dt^2}(r_1 - r_2) = -\frac{M_1 + M_2}{M_1 M_2} F(M_1 M_2, r_{12}) \frac{r_1 - r_2}{r_{12}}$$

(1.288)

and the motion of the two bodies relative to their centre of gravity is given by

$$\xi_1 = r_1 - \frac{1}{M_1 + M_2}(M_1 r_1 + M_2 r_2) = +\frac{M_2}{M_1 + M_2}(r_1 - r_2)$$

$$\xi_2 = r_2 - \frac{1}{M_1 + M_2}(M_1 r_1 + M_2 r_2) = +\frac{M_1}{M_1 + M_2}(r_1 - r_2)$$

(1.289)

Therefore

$$\frac{M_1 + M_1}{M_2} \xi_1 = -\frac{M_1 + M_1}{M_2} \xi_2 = (r_1 - r_2)$$

(1.290)

The centre of gravity moves in a straight line and therefore the two bodies will always find themselves in opposite directions along the straight line at opposite ends and dividing the line joining them in constant ratio $M2 : M1$. Hence

$$\frac{|\xi_1|}{M_2} = \frac{|\xi_2|}{M_1} = \frac{r_{12}}{M_1 + M_2} \tag{1.291}$$

Proposition *XXXI*. Theorem *II*

If two bodies of varying mass attract each other according to the G.R.F.F. *Matrix and revolve about the common centre of gravity, by the same forces may be described around either body unmoved a figure of similar and equal to the figures which the bodies so moving describe around each other.*

Consider two bodies, a planet (P) and the sun (S) of masses $M1$ and $M2$ revolving about their common centre of gravity. The equation governing the motion of both bodies will be given by

$$M_1 \frac{d^2}{dt^2}(r_1 - r_2) = -\frac{M_1 + M_2}{M_2} F(M_1, M_2; r_{12}) \frac{r_1 - r_2}{r_{12}} \tag{1.292}$$

The motion of p, mass M1, under the attraction of the G.R.F.F. force of S, mass $M2$, assumed to immovable is governed by

$$M_1 \frac{d^2 S}{dt^2} = -F(M_1, M_2; |S|) \frac{S}{|S|} \tag{1.293}$$

Where S denotes the radius vector joining s and p. The above equations show that the orbit of P round S under their mutual attraction may be formally taken as described by the nature of the G.R.F.F. field directed towards and away from S which is the same as when S is immovable but with a force enhanced by the factor $(M_1 + M_2)/M_2 = (S + P)/S$.

The acceleration of P towards and away from S in the revolving orbit is the same as the acceleration of p towards the immovable S enhanced by the factor $(S+P)/S$ It follows then that

$$M_1 \frac{d^2 \xi_1}{dt^2} = \frac{M_1 M_2}{M_1 + M_2} \frac{d^2}{dt^2}(r_1 - r_2) = -F\left(M_1, M_2; \frac{M_1 + M_2}{M_1 M_2}|\xi_1|\right)\frac{\xi_1}{|\xi_1|} \qquad (1.294)$$

then we will have

$$\xi_1 = \frac{M_2}{M_1 + M_2} X$$

$$\tau = t\left(\frac{M_1 + M_2}{M_2}\right)^{1/2} \qquad (1.295)$$

than we will have

$$M_1 \frac{d^2 X}{d\tau^2} = -F\left(M_1, M_2; |X|\right)\frac{X}{|X|} \qquad (1.296)$$

we will then have

$$S(t) = X(\tau) = \frac{M_1 + M_2}{M_2} \xi_1\left(t\sqrt{\frac{M_2}{M_1 + M_2}}\right) \qquad (1.297)$$

The orbit described by $M1$ under the influence of the G.R.F.F. of $M2$, to be immovable, is similar to the orbit described by $M1$ about the common centre of gravity C under their mutual interaction according to the G.R.F.F..

Cor. I Given two bodies whose forces are inversely proportional to the square of their distance, describe both round their common centre of gravity, and round each other, conic sections having their focus in the centre about from which these figures are described. And if such figures are described by the G.R.F.F. are inversely proportional to the square of the distance.

Cor. II Given two bodies whose interaction as determined by the G.R.F.F., are inversely proportional to the square of their distance, describe both round their common centre of gravity and each other, conic sections, having their focus in the

178

centre about which the figures will be described, the *G.R.F.F.*'s are inversely proportional to the square of their distance.

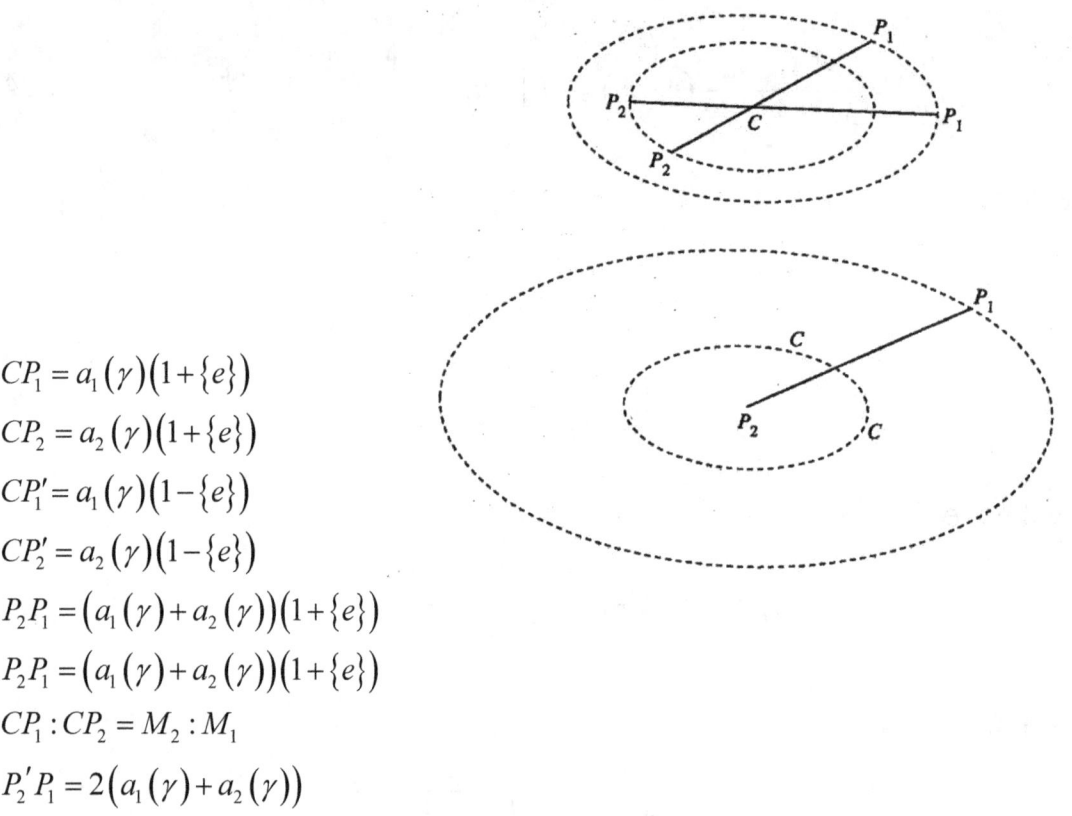

$$CP_1 = a_1(\gamma)(1+\{e\})$$

$$CP_2 = a_2(\gamma)(1+\{e\})$$

$$CP_1' = a_1(\gamma)(1-\{e\})$$

$$CP_2' = a_2(\gamma)(1-\{e\})$$

$$P_2P_1 = (a_1(\gamma)+a_2(\gamma))(1+\{e\})$$

$$P_2P_1 = (a_1(\gamma)+a_2(\gamma))(1+\{e\})$$

$$CP_1 : CP_2 = M_2 : M_1$$

$$P_2'P_1 = 2(a_1(\gamma)+a_2(\gamma))$$

Cor. *III.* Any two bodies revolving round their common center of gravity describe areas that are directly proportional to the times, that is by radii drawn to the center and each other.

Proposition *XXXII.* Theorem *III*

The periodic times of two bodies, denoted S and P, revolving round their common center of gravity C, is to the periodic time of one of the bodies P revolving round the other S remaining fixed, and describes a figure equal to those which the bodies describe about each other, as $\sqrt{S}$ is to $\sqrt{(S+P)}$

We write then

$$t = \left(\frac{M_2}{M_1 + M_2} \right)^{1/2} \tau : R = r_1 - r_2 \qquad (1.298)$$

we will then have for our equation (1.288)

$$M_1 \frac{d^2 R}{d\tau^2} = -F\left(M_1, M_2; |R|\right) \frac{R}{|R|} \qquad (1.299)$$

which is the same in form as the equation (1.293)

$$M_1 \frac{d^2 S}{dt^2} = -F\left(M_1, M_2; |S|\right) \frac{S}{|S|} \qquad (1.300)$$

governing the motion of M_1 about an immovable mass M_2. Then by comparing equations (1.288) and (1.299) we will have

(periodic time of M_1 revolving round M_2 under their gravitational relations / periodic time of M_1 about an immovable M_2 in a congruent orbit)

$$= \frac{t}{\tau} = \sqrt{\left[M_2 / \left(M_1 + M_2 \right) \right]} = \sqrt{S : \sqrt{(S+P)}}$$

Proposition *XXXIII*. Theorem *IV*

If two bodies S and P, are engaging in gravitational behavior according to the G.R.F.F., inversely proportional to the square of their distance, revolve about their common center of gravity: Then the principle axis of the ellipse in which either bodies, as P describes by this motion about the other body S, will be to the principle axis of the ellipse, which the body P may describe in the same periodic time about the other body S fixed, as to the sum of the two bodies to the first of two mean proportional's, between that sum and the other body S.

The G.R.F.F. may given as

$$F\left(M_1, M_2; r_{12}\right) = \frac{M_1 M_2}{r_{12}^2} \qquad (1.301)$$

For F, in the previous form, equations (1.292) and (1.293) will be given as

$$\frac{d^2 R}{dt^2} = -\left(M_1 + M_2\right)\frac{R}{|R|^3} \qquad (1.302)$$

and

$$\frac{d^2 S}{dt^2} = -\left(M_2\right)\frac{S}{|S|^3} \qquad (1.303)$$

The periodic times (periodic times in ellipses are as the 3/2 power of the greater axes), times for elliptical orbits, of semiaxis a_k and a_s we will have then

$$T_R = 2\pi\frac{a_R^{3/2}\left(\gamma\right)}{\left(M_1 + M_2\right)} : T_S = 2\pi\frac{a_S^{3/2}\left(\gamma\right)}{M_2^{1/2}} \qquad (1.304)$$

Hence if

$$T_R = T_S, \left(\frac{a_R^{3/2}\left(\gamma\right)}{\left(M_1 + M_2\right)}\right)^{1/2} = \left(\frac{a_S^{3/2}\left(\gamma\right)}{M_2^{1/2}}\right)^{1/2} \qquad (1.305)$$

or

$$a_R : a_S = \sqrt[3]{\left(M_1 + M_2\right)} : \sqrt[3]{M_2} \qquad (1.306)$$

while if

$$a_R = a_S, T_R : T_S = \sqrt{M_2} : \sqrt{\left(M_1 + M_2\right)} \qquad (1.307)$$

Proposition *XXXIV*. Theorem *V*

If two bodies attracting each other with the previously defined set of forces, and not otherwise disturbed by other forces and or moved in any other way, those motions will be the same as if they did not attract each other, but were both attracted with the same forces by a third body placed in their common centre of gravity, and the law of attracting forces will be the same in respect of the distance of the bodies from the common centre.

We will have

$$M_1 \frac{d^2\xi_1}{dt^2} = -F\left(M_1 M_2; \frac{M_1 + M_2}{M_2}|\xi_1|\right)\frac{\xi_1}{|\xi_1|}$$

$$M_2 \frac{d^2\xi_2}{dt^2} = -F\left(M_1 M_2; \frac{M_1 + M_2}{M_1}|\xi_2|\right)\frac{\xi_2}{|\xi_2|}$$

(1.308)

Because the law of attraction is inversely proportional to the square of the distance the above equation becomes

$$\frac{d^2\xi_1}{dt^2} = -M_2\left(\frac{M_2}{M_1 + M_2}\right)^2 \frac{\xi_1}{|\xi_1|^3}$$

(1.309)

the motion of $M1$ around the centre of gravity C is the same if C was an immovable mass,

$$M_2\left(\frac{M_2}{M_1 + M_2}\right)^2$$

(1.310)

and $M2$ was absent. We will then have

$$\xi_1 = \frac{M_2}{M_1 + M_2}\eta$$

$$\xi_2 = \frac{M_1}{M_1 + M_2}\eta$$

(1.311)

than both equations will take the form

$$\frac{d^2\eta}{dt'} = -\frac{M_1 + M_2}{M_1 M_2} F(M_1, M_2; |\eta|)\frac{\eta}{|\eta|}$$

(1.312)

Newly formed initial-value problems

Proposition *XXXV*. Theorem *VI*

We seek to determine the motions of two bodies which attract each other with devised forces inversely proportional to the squares of the distance between them, and are all let fall from given locations.

In Proposition *XVI* , the solution of the problem of rectilinear motion descent of a body, originally at rest, under the influence of an stationary center of force, is with a force inversely proportionally to the square of the distance as given. In this proposition it can be shown that the solution can be adapted to the motion of two bodies, at rest, influencing each other with the same inverse-square law of force.

The bodies, like in the last theorem, will be moved in the same manner as if they were affected by the presence of a third body, in place in the common center of gravity; and the center being fixed at the start of the motion, then by Cor. *IV* of the laws of motion, will then always be fixed. The motions of the bodies will be determined in the same manner as if they were affected by a force related to the center of gravity, and we will have then the motions of the bodies influencing each other by their presence.

Proposition *XXXVI*. Theorem *VII*

We seek to determine the motions of two bodies affecting each other according to the laws of the G.R.F.F.'s, inversely proportional to the squares of the distance, and going away from given places in given directions with given velocities.

In an earlier proposition, in the section on conic sections, given the gravitational force previously defined that determines the line of motion, we have shown how the motion of a body, under the influence of the *G.R.F.F.*, of an immovable center with a force inversely proportional to the square of the their distance, follows an initially assigned position and velocity then found.

The motion of bodies at the start being given, there is likewise the uniform motion of the common center of gravity, as the frame of reference. Then the bodies motion will be altered relative to the motion of the center of gravity, as if it were at rest, and as if the bodies did not affect each other, but were affected by the presence of a third body positioned in that center. The motion then in regard to this movable frame of reference in a given direction, with a given velocity, and affected by the *G.R.F.F.*, is to be determined by Prop. *IX* and *XXVI* , and at the same time will be obtained by the motion of the other around that same center.

New solution to the many-body problem

Proposition *XXXVII*.   Theorem *VIII*

Given the known forces with which bodies attract each other, to increase in a simple ratio of their distances from the centers; we seek to find the motions of several bodies amongst themselves.

Given than N mass points, $M_1, M_2, ..., M_N$, the mass M_i being attracted by the mass M_j with a force

$$-KM_i, M_j \left(r_i - r_j \right)$$
$$(i \neq j)$$

(1.313)

K is some positive constant, we write then

$$-K_i K_j \left(r_i - r_j \right)$$
$$(i \neq j)$$

(1.314)

where K is some positive constant. The relevant equations of motion are than

$$M_i \frac{d^2 r_i}{dt^2} = -K^2 \sum_{\substack{j=1 \\ j \neq i}}^{N} M_i, M_j \left(r_i - r_j \right)$$

$$(i = 1, ..., N)$$

(1.315)

These will than admit the integral

$$\sum_{i=1}^{N} M_i r_i = \left(M_1 + M_2 + ... + M_N \right) Vt + A$$

(1.316)

where V and A are constant vectors. In a frame of reference where the centre of gravity is at rest at the origin, we will then have

$$\sum_{j=1}^{N} M_i r_i = 0$$

$$(V = 0, A = 0)$$

(1.317)

In this frame, the equations of motion will be

$$\frac{d^2 r_i}{dt^2} = -K^2 \left(\sum_{j \neq i}^{N} M_j \right) r_i + K^2 \sum_{j \neq i}^{N} M_j r_j = -K^2 \left(\sum_{j \neq i}^{N} M_j \right) - K^2 M_i r_i$$

$$(i = 1, ..., N)$$

(1.318)

Than

$$\frac{d^2 r_i}{dt^2} = -K^2 \left(\sum_{j \neq i}^{N} M_j \right) r$$

(1.319)

If follows than

$$r_i = A_i e^{\pm i\sigma t} \quad (A_i = \text{constant vector})$$

(1.320)

where

$$\sigma = K \sqrt{\sum_{j \neq i}^{N} M_j} = K \sqrt{} \text{ (mass of the system)}$$

(1.321)

Consider two bodies T and L, of masses $M1$ and $M2$, attracting each other with the previously defined force. The equations governing their motions will then be

185

$$\frac{d^2 r_2}{dt^2} = -K^2 M_1 \left(r_2 - r_1 \right) \tag{1.322}$$

and

$$\frac{d^2 r_1}{dt^2} = -K^2 M_2 \left(r_1 - r_2 \right) \tag{1.323}$$

In the frame of reference in which the centre of gravity D, of the masses is at rest, we will have then

$$D = M_1 r_1 + M_2 r_2 = 0 \tag{1.324}$$

in this frame the equations are

$$\frac{d^2 r_1}{dt^2} = -K^2 \left(M_1 + M_2 \right) r_1$$

$$\frac{d^2 r_2}{dt^2} = -K^2 \left(M_1 + M_2 \right) r_2 \tag{1.325}$$

both masses describe elliptical orbits with the same period about D as centre.

Let a third body S, of mass $M3$, be introduced and such that it gravitationally affects T and L. In turn it is attracted by the other bodies. By including the attraction by S, the motions of T and L, in the same frame as before, will be governed by the following equations

$$\frac{d^2 r_1}{dt^2} = -K^2 (M_1 + M_2) r_1 - K^2 M_3 (r_1 - r_3) = -K^2 (M_1 + M_2 + M_3) r_1 + k^2 M_3 r_3$$

$$\frac{d^2 r_2}{dt^2} = -K^2 (M_1 + M_2) r_2 - K^2 M_3 (r_2 - r_3) = -K^2 (M_1 + M_2 + M_3) r_2 + k^2 M_3 r_3 \tag{1.326}$$

Therefore $M1$ and $M2$, aside from executing elliptical orbits about D, quicker than in the absence of $M3$, are moved parallel in the direction r_3 with same acceleration $K^2 M_3 r_3$.

By considering the motion of S, the equation governing its motion will be

$$\frac{d^2 r_3}{dt^2} = -K^2 M_1 (r_3 - r_1) - K^2 M_2 (r_3 - r_2)$$

$$= -K^2 (M_1 + M_2) r_3 + K^2 (M_1 r_1 + M_2 r_2)$$

(1.327)

or also as

$$\frac{d^2 r_3}{dt^2} = -K^2 (M_1 + M_2) r_3 = -K^2 (M_1 + M_2 + M_3) r_3 + k^2 M_3 r_3$$

(1.328)

Now by changing the frame of reference, from the one in which the centre of gravity of $M1$ and $M2$ is at rest in relation to the one in which the centre of gravity of $M1$, $M2$, and $M3$, is at rest, we will have then

$$M_1 r_1 + M_2 r_2 + M_3 r_3 = 0$$

(1.329)

that is, to a frame which is which is accelerated in regard to a frame of reference by $-KM_3 r_3$, the equations governing the motions of the three bodies will given as

$$\frac{d^2 r_i}{dt^2} = -K (M_1 + M_2 + M_3) r_i$$

(1.330)

Framed-dependent form for orbital eccentricity

Before we start we need to define the values for the frame-dependent variables. This can be given as

$$\pm a \rightarrow \pm a(\gamma) = \pm a \sqrt{\left(1 - \frac{V^2}{\alpha^2}\right)}$$

(1.331)

where α is an arbitrarily large number. Where

$$\gamma = \sqrt{\left(1 - \frac{V^2}{\alpha^2}\right)}$$

(1.332)

And also

$$e \rightarrow [e] = \frac{c}{a(\gamma)} \tag{1.333}$$

where

$$c = \sqrt{\left\{ \left(a(\gamma)\right)^2 - \left(\frac{b}{(\gamma)}\right)^2 \right\}} \tag{1.334}$$

We place the foci on the $x-$axis at $F_1(-c,0)$ and $F_2(c,0)$, so that the origin is halfway between them. The sum of the distance from a given point on the ellipse will be the foci $2a$. Then if $P(x,y)$ is any point on the ellipse we will have

$$d(P,F_1) + d(P,F_2) = 2a \tag{1.335}$$

According to the Distance Formula

$$\sqrt{(x+c)^2 + y^2} + \sqrt{(x-c)^2 + y^2} = 2a \tag{1.336}$$

or

$$\sqrt{(x-c)^2 + y^2} = 2a - \sqrt{(x+c)^2 + y^2} \tag{1.337}$$

then squaring each side and expanding we will get

$$x^2 - 2cx + c^2 + y^2 = 4a^2 - 4a\sqrt{(x+c)^2 + y^2} + \left(x^2 + 2cx + c^2 + y^2\right) \tag{1.338}$$

which simplifies to

$$4a\sqrt{(x+c)^2 + y^2} = 4a^2 + 4cx \tag{1.339}$$

then dividing each side by 4 and squaring again we get

$$a^2\left[(x+c)^2 + y^2\right] = \left(a^2 + cx\right)^2$$
$$a^2x^2 + 2a^2cx + a^2c^2 + a^2y^2 = a^4 + 2a^2cx + c^2x^2 \tag{1.340}$$
$$\left(a^2 - c^2\right)x^2 + a^2y^2 = a^2\left(a^2 - c^2\right)$$

Kepler's laws of planetary motion

First Law:

An ellipse can be given by the formula

$$r = \frac{p}{1 + \{e\} \cos \theta} \qquad (1.341)$$

p is the semi-latus rectum and $\{e\}$ is the newly developed eccentricity. And r is the distance from the sun to the planet and θ is the angle to the planet's current position from its closest approach. Both are given in polar coordinates.

Third Law:

The square of the orbital period of a planet is directly proportional to the cube of the semi-major axis of its orbit. This can be represented by the following equation

$$P^2 / a(\gamma)^3 \qquad (1.342)$$

11

The method of the variation of the elements of a Kepler orbit and a new formation of lunar theory

Introduction

We seek to express the variations, with time, of the elements of the Kepler orbit under the action of a force devised as the G.R.F.F.*, elements that otherwise would have been constants.*

Basic equations, definitions, and coordinate system

$F_{\alpha\beta}$ will be given as

$$\Gamma_{ij_\alpha} = \pm\left\{\pm\sum(\sin\alpha TP \pm \cos\alpha TP) \pm \sin\phi SP + \sum(\pm\sin\beta KR \pm \cos\beta KR)\right\} \quad (1.343)$$

or

$$\Gamma_{(ij)_\beta} = \pm\left\{\pm\sum(\sin\alpha TP \pm \cos\alpha TP) \pm \cos\phi SP + \sum(\pm\sin\beta KR \pm \cos\beta KR)\right\} \quad (1.344)$$

Suppose at some point P, where the position r and the velocity $v = \dfrac{dr}{dt}$ are given, the perturbing forces cease. The body will than move in an ellipse defined by the nature of the above-mentioned field. This is also the osculating ellipse at P. It is a

Kepler orbit with at that point the position and velocity as the true orbit. This will be a variation of the initial variation of the elements. In terms of the newly devised field, our equation depicting velocity will be given as

$$\frac{dv}{dt} = -\frac{u}{r^3} r + F_{(ij)}$$

$$\rightarrow -\frac{u}{r^3} r \pm \left\{ \pm \sin \phi SP \pm \sum (\sin \alpha TP \pm \cos \alpha TP) + \sum \left(\pm \sin \beta KR \pm \cos \beta KR \right) \right\}$$

(1.345)

we define the vectors, h, e, (no longer constants) by the following equations:

$$\left[h \right] = \omega \times m$$

$$u \left(\frac{r}{r} + e \right) = v \times \left[h \right]$$

$$e \cdot v = -\frac{1}{r} r \cdot v$$

(1.346)

we will then have

$$\frac{d}{dt} \frac{r}{r} = \frac{\left[h \right] \times r}{r^3}$$

(1.347)

and

$$\frac{dh}{dt} = r \times \left(-u \frac{r}{r^3} + F_{(ij)_\alpha} \right) = r \times F_{(ij)_\alpha}$$

(1.348)

and from equations (1.346) and (1.347) we will have

$$u \left(\frac{\left[h \right] \times r}{r^3} + \frac{de}{dt} \right) = \frac{dv}{dt} \times \left[h \right] + v \times \frac{d \left[h \right]}{dt} = \left(-u \frac{r}{r^3} + F_{(ij)_\alpha} \right) \times \left[h \right] + v \times \left(r \times F_{(ij)_\alpha} \right)$$

(1.349)

or

$$u \frac{de}{dt} = F_{(ij)_\alpha} \times \left[h \right] + v \times \left(r \times F_{(ij)_\alpha} \right)$$

(1.350)

The coordinate System

$i, j,$ and k are the unit vectors along the axis of the rectangular system of coordinates, in respect to the ecliptic plane. The direction of the vernal equinox and the ascending node are random and not defined in terms of the $G.R.F.F.$. The angular momentum vector, defined in terms of the orbiting body, is a direct result of the $G.R.F.F.$.

The perturbed orbit, it's magnitude, is proportional to the diameter of the third body or mass, the perturbing force in question, and its distance from the orbiting body in question.

It is inversely proportional to the distance from both orbiting bodies in question.

We will then

$$k \pm \left\{ \sin \phi u + \left(\left\{ \sum (-\sin + \cos) AB \right\} + \left\{ \sum (\sin + (-\cos)) TP \right\} \right) \frac{Mm}{r^2} \right\}$$
$$= k \pm \left\{ \sin \phi u + \left(\left\{ \sum (-\sin + \cos) v \right\} + \left\{ \sum (\sin + (-\cos)) \mu \right\} \right) \frac{Mm}{r^2} \right\}$$

(1.351)

Here we choose a right-handed coordinate system, defined by the unit vectors

$$l_i, l_{jk_\alpha}, l_{jk_\beta}$$

(1.352)

along the directions

$$i, jk_\alpha, jk_\beta$$

(1.353)

of magnitudes

$$i = |i|, jk_\alpha = |jk|, jk_\beta = |jk|$$

(1.354)

along these vectors we will have

$$r \times \alpha = r^2 h, \alpha \times h = h^2 r, \alpha \cdot r$$

(1.355)

we will than resolve the force F along the unit vectors with the components

$$F_i, F_{jk_\alpha}, F_{jk\beta} \rightarrow F_i, F_{jk_\alpha} = F_{jk\beta} \qquad (1.356)$$

and for the generalized form of the gravitational perturbation we will have the following equalities

$$\Gamma_{i1} = \pm\left\{\pm\sin\phi SP \pm \sum(\sin\alpha TP \pm \cos\alpha TP) + \sum(\pm\sin\beta KR \pm \cos\beta KR)\left(\frac{A_1 A_2}{r^2}\right)\right\}_h$$

$$\Gamma_{i2} \pm\left\{\pm\cos\phi SP \pm \sum(\sin\alpha TP \pm \cos\alpha TP) + \sum(\pm\sin\beta KR \pm \cos\beta KR)\left(\frac{A_1 A_2}{r^2}\right)\right\}_h$$

$$\Gamma_{jk_\alpha 1} = \pm\left\{\pm\sin\phi SP \pm \sum(\sin\alpha TP \pm \cos\alpha TP) + \sum(\pm\sin\beta KR \pm \cos\beta KR)\left(\frac{A_1 A_2}{r^2}\right)\right\}_r \quad (1.357)$$

$$\Gamma_{jk_\alpha 2} \pm\left\{\pm\cos\phi SP \pm \sum(\sin\alpha TP \pm \cos\alpha TP) + \sum(\pm\sin\beta KR \pm \cos\beta KR)\left(\frac{A_1 A_2}{r^2}\right)\right\}_r$$

$$\Gamma_{jk_\beta 1} \pm\left\{\pm\sin\phi SP \pm \sum(\sin\alpha TP \pm \cos\alpha TP) + \sum(\pm\sin\beta KR \pm \cos\beta KR)\left(\frac{A_1 A_2}{r^2}\right)\right\}_\alpha$$

$$\Gamma_{jk_\beta 2} \pm\left\{\pm\cos\phi SP \pm \sum(\sin\alpha TP \pm \cos\alpha TP) + \sum(\pm\sin\beta KR \pm \cos\beta KR)\left(\frac{A_1 A_2}{r^2}\right)\right\}_\alpha$$

where A_1 and A_2 are the respective total surface area in three-dimensional volumes.

i, j, k : unit vectors along the axes of a rectangular system of coordinates defined with respect to the ecliptic plane;

i : direction of the vernal equinox;

v : direction of the ascending node;

h : angular momentum vector;

e : Lenz vector

$\Omega = \angle(v, i)$;

$\omega = \angle(v, e); \omega + \varphi = v$;

$\varphi = \angle(r, e)$

$$g = \left(h \times v \right);$$

$\iota = \angle \left(h, k \right) =$ inclination of the orbital plane to the ecliptic.

Let the velocity v resolved along the directions of r and α be

$$v = Ar + B\alpha \tag{1.358}$$

where A and B are to be determined. By this definition than we will have

$$\left[h \right] = \omega \times m = B\left(r \times \alpha \right) = Br^2 \left[h \right] \tag{1.359}$$

therefore

$$B = r^{-2} \tag{1.360}$$

The Variation of the Elements

$\left[h \right]$: angular momentum vector;

$\iota = \angle \left(\left[h \right], k \right) =$ inclination of the orbital plane to the ecliptic;

Ω: the angle of the ascending node in the equatorial plane

e: the Lenz vector

ω: the inclination of e to the direction of v;

a: the semimajor axis; and $n = 2\pi \, / \, period$

F_{ij} along the direction r

we will have

(a) Variation of h

Given

$$\frac{dh}{dt} = r \times F_{ij_\alpha} \qquad (1.361)$$

therefore

$$h \cdot \frac{d[h]}{dt} = r[h] = rF_{ij} = r\left(F_i \frac{r}{r} + F_{jk} \frac{r_{jk}}{r_{jk}} + F_{lm} \frac{r_{lm}}{r_{lm}} \right) \qquad (1.362)$$

(b) Variation of I

Since

$$[h] \cdot k = [h] \cos l \qquad (1.363)$$

we will have

$$\frac{dh}{dt} \cdot k = \frac{dh}{dt} \cos \iota = rF_{ij_\alpha} \cos \iota \qquad (1.364)$$

than by equation (1.361)

$$\frac{dh}{dt} \cdot k = r \times F_{ij} \cdot k = r \times F_{ij} \cos l = r \times \left(F_i \frac{r}{r} + F_{jk_\alpha} \frac{r_{jk_\alpha}}{r_{jk_\alpha}} + F_{jk_\beta} \frac{r_{jk_\beta}}{r_{jk_\beta}} \right) \cos l \qquad (1.365)$$

(d) Variation of e

We begin with

$$u \frac{de}{dt} = [F_{\alpha\beta}] + [h] - (r \cdot v)[F_{\alpha\beta}] + ([F_{\alpha\beta}] \cdot v) r \qquad (1.366)$$

By resolving F into its components and substituting for v from prior equations we obtain successively

$$\mu \frac{de}{dt} = \left(F_i + F_{jk} + F_{lm} \right) \frac{r \times [h]}{r} - Ar^2 \left(F_i \frac{r}{r} + F_{jk} \frac{r_{jk}}{r_{jk}} + F_{lm} \frac{r_{lm}}{r_{lm}} \right)$$

$$+ \left(Ar \left(F_i \frac{r}{r} + F_{jk} \frac{r_{jk}}{r_{jk}} + F_{lm} \frac{r_{lm}}{r_{lm}} \right) \right) r$$

(1.367)

Multiplying this equation scalarly be e and remembering that it is orthogonal to h and inclined to the direction of r by the angle φ we will find

$$\mu \frac{de}{dt} \cdot e = \frac{1}{r} \left(F_i (\sin \varphi) + F_{jk_\alpha} (\sin \varphi + \cos \varphi)_j + F_{jk_\alpha} (\sin \varphi + \cos \varphi)_k \right) ehr$$

$$= he \frac{1}{r} \left(F_i (\sin \varphi) + F_{jk_\alpha} (\sin \varphi + \cos \varphi)_j + F_{jk_\alpha} (\sin \varphi + \cos \varphi)_k \right)$$

(1.368)

or

$$\frac{\mu}{h} \frac{de}{dt} = \left(F_i (\sin \varphi) + F_{jk_\alpha} (\sin \varphi + \cos \varphi)_j + F_{jk_\alpha} (\sin \varphi + \cos \varphi)_k \right)$$

(1.369)

(e) Variation of $\omega = \angle (v, e)$

Since

$$(k, e) = e \sin l \cos \omega$$

(1.370)

than

$$k \cdot \frac{de}{dt} = \sin \iota \sin \omega \left(\frac{h}{\mu} \right) \left(F_i (\sin \varphi) + F_{jk_\alpha} (\sin \varphi + \cos \varphi)_j + F_{jk_\alpha} (\sin \varphi + \cos \varphi)_k \right)$$

$$+ e \sin \iota \cos \omega \frac{d\omega}{dt}$$

(1.371)

$$k \cdot \frac{de}{dt} = \frac{k}{u} \cdot \left(\frac{1}{r} \left(F_i (\sin \varphi) + F_{jk_\alpha} (\sin \varphi + \cos \varphi)_j + F_{jk_\alpha} (\sin \varphi + \cos \varphi)_k \right) \right)$$

$$= \left(\frac{1}{r} \left(F_i (\sin \varphi) + F_{jk_\alpha} (\sin \varphi + \cos \varphi)_j + F_{jk_\alpha} (\sin \varphi + \cos \varphi)_k \right) \right) hr \sin \iota \cos (\omega + \varphi)$$

(1.372)

(f) Constructing the values for the ellipse

We start with the equation

$$(k \cdot e) = \{e\} \sin \iota \sin \omega \qquad (1.373)$$

Were $e \to \gamma$ is a frame-dependent phenomenon, then we will have

$$e \frac{d\omega}{dt} = -\left(\frac{[h]}{\mu} \cos\varphi \right) \left(F_i (\sin\varphi) + F_{jk_\alpha} (\sin\varphi + \cos\varphi)_j + F_{jk_\alpha} (\sin\varphi + \cos\varphi)_k \right) \qquad (1.374)$$

then for the osculating ellipse we will have the set of equations

$$r = \frac{l}{1 + e\cos\varphi} = \frac{[h]^2}{u} \frac{l}{1 + e\cos\varphi} = \frac{[h]^2}{1 + e\cos\varphi} = \frac{a(\gamma)(1 - e^2)}{1 + e\cos\varphi} \qquad (1.375)$$

If u denotes the eccentric anomaly, then we will have

$$r = a(1 - e\cos u) \qquad (1.376)$$

also given as

$$e\cos u = 1 - \frac{r}{a} = 1 - \frac{1 - e^2}{1 + e\cos u} = e \frac{e + \cos u}{1 + e\cos u} \qquad (1.377)$$

(g) Variation of a and n

If we vary a and n we will then have

$$2\frac{h_t}{h} = \frac{a_t}{a} - \frac{2ee_t}{1 - e^2} = \frac{a_t}{a} - \frac{2uaee_t}{h^2} \qquad (1.378)$$

or differently as

$$\frac{1}{2}[h]\frac{a_{,t}(\gamma)}{a(\gamma)} = [h_{,t}] + \frac{\mu a(\gamma)}{[h]}(ee_t)$$

$$= ae\left(F_i(\sin\varphi) + F_{jk_\alpha}(\sin\varphi+\cos\varphi)_j + F_{jk_\alpha}(\sin\varphi+\cos\varphi)_k\right)$$

(1.379)

and

$$\frac{1}{2}[h]\frac{a_t(\gamma)}{a(\gamma)} = [h]_t + \frac{ua(\gamma)}{[h]}\{e\}\{e\}_t = r\left[F_{\alpha\beta}\right] + a(\gamma)\{e\}\{[F_u]\sin\varphi i\}$$

$$= r\left(F_u\frac{r}{r} + F_{uv_\alpha}\frac{r_{ij}}{r_{ij}} + F_{uv_\beta}\frac{r_{ij}}{r_{ij}}\right) + a(\gamma)\{e\}\{[F_u]\sin\varphi i\}$$

The coefficients combine to make

$$r\left\{1 + \frac{e}{1-e^2}\left[e(1+\cos^2\varphi) + 2\cos\varphi\right]\right\} = \frac{h^2 a}{ru}$$

(1.380)

(h) Variation of Keplers equations

The variation of the equation r gives us

$$r_t = (1-e\cos u)a_t + (ae\sin u)u_t - (a\cos u)e_t$$

(1.381)

Than

$$r_t = v = \frac{\sqrt{(\mu a)}}{r}e\sin u$$

(1.382)

Where $\mu = \left(F_{ij} \oplus F_{kl}\right)\frac{Mm}{r^2}$

Then

$$\frac{\sqrt{(\mu a)}}{r}e\sin u = \frac{r}{a}a_t + (ae\sin u)u_t - (a\cos u)e_t$$

(1.383)

given the time *(t-T)* passed after passing the vernal equinox at time *T* we will have

198

$$n(t-T) = u - e\sin u \tag{1.384}$$

than variation of this equation will give us

$$n_t(t-T) + n(1-T_t) = (1-e\cos u)u_t - (\sin u)e_t \tag{1.385}$$

than removing u_t from equation (1.383) we will then have

$$\left(\frac{\mu}{a}\right)^{1/2}\left[\frac{3}{2}\frac{a_t}{a}(t-T)+T_t\right]e\sin u = \frac{r^2}{a^2}a_t + (ae\sin^2 u - r\cos u)e_t \tag{1.386}$$

than making some simplifications we will then have

$$\left(\frac{\mu}{a}\right)^{1/2}\left[\frac{3}{2}\frac{a_t}{a}(t-T)+T_t\right]e\sin u = \frac{r^2}{a^2}a_t - (r\cos\varphi)e_t \tag{1.387}$$

12

The three-body problem: Lunar theory part 1

Introduction

Here is the first part of the Lunar theory reavaluated, ther three-body problem.

Proposition *XXXVIII*. Theorem *I*

Bodies, whose forces decrease as the direct square of their distances from their centres, will move among themselves in ellipses; and also by radii drawn to the foci may describe areas very nearly proportional to the times.

Case 1. Consider a larger body M, about which several smaller bodies, m_1, m_2, and so on revolve around the larger body. The centre of gravity of these smaller bodies should either be at rest or move uniformly in a right line. If the smaller bodies are sufficiently smaller, than the centre of gravity will not be sensibly different from the location of M, which may be either at rest or moving uniformly forward in a right line and about which the lesser bodies will revolve. If the mutual attractions are sufficiently smaller compared to the attraction of the larger body than the smaller bodies will continue to revolve about M in ellipses describing equal areas in time.

Case 2. Consider the same system as in case 1. Except now the attractive force between m_1 and m_2, the smaller bodies, is strong enough that they describe elliptical orbits about the common centre of gravity M. If the larger dass M is far enough away from the distance between the smaller bodies, is small compared to the distance MA, then will attract the smaller bodies along nearly parallel lines and subject them to equal parallel accelerations. Since equal and parallel accelerations will not affect their relative motions, it follows that M will act on the smaller bodies as if the two together were actually one body. Than following this the centre of gravity the smaller bodies will describe about M a parabola or a hyperbola.

Proposition *XXXIX.* Theorem *II*

If three bodies whose forces decrease as the square of the distances, attract each other by the fundamental G.R.F.F.*; and these accelerative forces of any two towards the third be between themselves inversely as the square of the distances; and the smaller bodies revolve about the greater body than the interior of the two revolving bodies will by radii drawn to the innermost and greatest body. They describe round that body areas more proportional to the times, and a figure more approaching to that of an ellipse having its focus at the point of intersection of the radii. Let the larger body be agitated by those attractions as determined by the said field, than it would do if that greater body were not attracted at all by the smaller, but remained at rest; or than it would do if that larger body were much less or much more attracted, or very much or very much less agitated by those previously defined attractions.*

Case. 1 We are given three bodies, T, P, and S. Under their mutual interaction as defined by the G.R.F.F., we consider the perturbation of the elliptical orbit described by a smaller body about the larger body P, about the largest body T, and by another smaller body S. S is considered a smaller body in the sense that its influence on the motion of P describes about T.

Proposition *XXXX.* Theorem *III*

Every body, by a radius drawn to the center of another body, describes areas about that center proportional to the times, is moved so perpetually to the center by two equal but distributed forces moving directly opposite of each other, is as well drawn to the surface of the bodies as illustrated by a centripetal force.

Both bodies are positioned by external forces to move towards each other, but by consequence of the two forces are to be moved in orbital revolutions of each other.

Let L be one body and T the other body. Both are urged by forces, characterized as collective moving particles both microscopic and macroscopic. The forces exert force on the opposing sides of both bodies. Each line of force characterized by the mass of moving particles, is moved slightly aside from the center of the body by a factor of π. This factor is equal to both bodies at all times. Both bodies are issued to move towards each other, but by the nature of the described forces move in orbital relation to each other. The magnitude of the force and the ratio of time and areas described by the orbiting body, is directly proportional to the mass of body emitting the particles, thus describing the amount of particles emitted in relation to the described area, and the mass of the body revolving. The magnitude of the gravitational field is in such ratio as distance from the emitting body in question and it's mass. The closer the stronger the field in question. The reason for this is that the closer the orbiting body is to the larger body in question, the greater the number of particles strike the surface of the body in orbital motion, in defined periods or epochs of time. The force are never exactly parallel to the each other, but are deviated by a small factor of π. Because the opposing forces are equal, they will cancel any linear motion, and due to the fact that they do not strike the object directly in the center of the object, the object is required to move with angular direction. Because the particles strike the object at the nearly exact same location throughout its motion, it will, perpetually so, move in a consistent orbital motion. Each force is general as a whole force but is also compounded by a collection of smaller forces, or particles. If we were to draw a radius from one body to another, the outside forces will push both objects together until they are close enough were the magnitude on both sides of both objects is equal forcing them to move than within a described orbital path, revolving about each other. Again the magnitude of the force depends on the mass of internal energy of the emitting body. The closer to the object the stronger the gravitational field.

Section II:

Let the time, with respect to a singular revolving body, be split into separate but equal intervals. Here we will describe the gravitational field. Let a body be positioned to move in a rectilinear motion from point to A to point B. If a force were to be exerted on the object at the same time it were to move from point A to point B, let a force moving directly to the center of the $G.R.F.F.$, denoted as S, be impressed upon the moving object. As the body begins to move from along it's predesigned path, the trajectory AB, it at the same time, will be moved from A to s, along the path As. Now just based on this it would continue to move along a new rectilinear path from A to s, or As. Another force is then impressed upon the same moving object. As the object is moving from A to s, the second force, equal in strength and moving in exactly the opposite direction, is impressed upon the same object positioned from A to B, or AB. The moving object will ultimately, in its motion, bisect both B and s and moving along a predesigned path towards P, in an arc motion whose sine value will remain equal throughout. Let us deem this setup in time interval T, as T_1. In time interval Tb, will have the second portion of the $G.R.F.F.$, deemed T_2. This is equal in all values to T_1. This will continue onwards throughout its orbital motion. Than all the other lines in succession to the initial line will be equal. Each line drawn from the center S to A, B, and so on will be equal in length for non-elliptical orbits. Once an eccentricity is factored in, the lengths themselves will ultimately change proportionally. Again for general circular orbits, the lengths and the areas circumscribed by the revolutions of the revolving body will be equal and in direct ratio of each other. Each two successive lines drawn as rectilinear lines will be superimposed onto each other as they run parallel to each other. For regular orbits each successive triangle drawn by the revolving body according to the generalized laws of the gravitational field, will be equal in area. In a space empty of resistance, the force drawn through the center of the revolving object, perpendicular to its surface, to the center of the field, one can draw right lines that are tangent to the orbital tendency of the path of revolution.

203

Proposition *XXXXI*. Theorem *IV*

We replace the superficial centripetal force with the newly developed G.R.F.F. The bodies elliptical motions described by a tendency towards the center is defined by the energy field by two equable forces.

They tend in the same system, to the same center described by the same square of the arcs, in equal times, in ratio to the radii of the ellipses. The square of the arcs describes the magnitude of the *G.R.F.F.* by the revolutionary momentum of the orbiting bodies in relation to the energy of the *G.R.F.F.* in terms of distance as the value of the square. The closer the object is to the center to the system, the greater the magnitude of the field, realized by the greater number of particles that exert radiation pressure. The greater but equal pressure exerted on the object, from both sides, the greater the slippage and thus revolutionary momentum. Regardless of the distance of the revolving object from the center of the system, the area described by that object is the same regardless. That is if we were to draw a radii from the revolving object, at one position, at time A, to the center and then another radii to the same center at time B, then the area described by that object, regardless of its distance from the same center, the areas described will likewise be the same. The revolving object will always be positioned directly perpendicular to the center of the system. The arcs described in equal times will have the same area regardless of the distance and thus the ratio in squares. The versed sines, the angle of arcs described, is the tendency of the revolving object to be forced towards and away from the center by a constant of acceleration.

Given a body, denoted as A and another body denoted as B, we can draw a radius connecting both bodies in a straight line. Let us say that the smaller body revolves around the larger body, or both revolve around each other, assuming equal masses, than the area of the triangles drawn connecting the length of the arc drawn from its path along this arc, and a line drawn from the center of the system that connect both points of the path of revolution, (between those two points along with arc) are in direct ratio of law of inverses. Both forces of propulsion for both or singular bodies is defined by the *G.R.F.F.*. We replace the fictitious centripetal force defined according to the inverse square relation, given by the *G.R.F.F.*.

Proposition *XXXXII*.   Theorem *V*

We draw a radius from one body, its center, to the center of another body, the force that binds them in a circular and orbiting motion, is defined so by the G.R.F.F..

By this force, each body is propelled to move towards and away from the other body. The areas described by the orbiting bodies about a common center is proportional to the times or amount of time needed to cover an area in relation to the lengths of the circumscribed arc and the lines drawn to the center of the system. Let there be two bodies, T and S. Assuming the body T to be larger than S, so that the orbiting body in question would be just the singular one. The relation between the bodies in question, in terms of gravitational behavior, will be defined by the *G.R.F.F.* Again we are just focused on circular motion not the existing elliptical motion found in all orbiting bodies.

Proposition *XXXXIII*.   Theorem *VI*

The combination of both the tendency of an object to be attracted to another object and be repelled at the same time, defines the G.R.F.F..

The inverse square law implies that the farther away the revolving object is from the center of the system the greater the length of the arc is ratio to its time, compounded proportionally. The farther away, equal on both sides, the greater the radiation pressure and thus the total number of particles striking the surface of the revolving body. The angles, or versed sines, of the revolving body is the same regardless of the distance or magnitude of the *G.R.F.F.*. The greater the range of the arc, with respect to the same area of the triangle of the body's motion, the greater the length of the length of the radii. The square of the arc relates directly to the diameter of the circle. The greater the value of the square, the greater the diameter. Now with respect to the value of the field, the greater the value of the square in relation to the mass of the center of the field, the weaker the strength of the field, with respect to the increase in the diameter of the circle. The length of the arc depends on the velocity of the revolving body in relation to the length of the radii or its square. The body's orbiting velocity depends on the strength of the field. The field depends on the number of particles that strike its surface over the definitive norm or epoch of time along the sine of the arc, where the opposing forces have the same trigonometric value (sine and negative sine or cosine and negative cosine). Each line represents a stream of particles that act as a constant force, and the closer the object is to the center of the system the greater the number of lines of force. The lines deviate slightly from the center of the revolving body's surface by a defined and constant value of π. The greater the radii the greater the times. The times are in ratio compounded of velocities inversely, which is defined by greater times lesser velocity. The $G.R.F.F.$ is in an inverse ratio with respect to the velocities and a direct ratio of the radii and in duplicate ratio of the periods, in terms of time, inversely. The $G.R.F.F.$ is always in an inverse ratio with respect to the product of the two masses in question. The velocities are in direct ratio to the radii. The greater the periodic times the greater the radii and the lesser, in proportion, the velocities. The $G.R.F.F.$ is in direct ratio to the radii which correlates to the velocities and periodic times.

Proposition *XXXXIV*. Theorem *VII*

Given a circle, the arc is defined by nature of the $G.R.F.F.$

The versed sine, or angle, of the arc is in direct ratio to the strength of the $G.R.F.F.$. The verse sine of the arc is drawn by bisecting a chord that passes through the center of the system. The $G.R.F.F.$ force positioned in the center of the arc, is equal to the versed sine and the square of the time inversely. The greater the time the greater the diameter and angle of revolution. Let a body revolve about the point S at the center of the system. Given the $G.R.F.F.$ and its ratios, the body describes an arc

PQR and *TRM*, which is touched at point *Q* by a tangent line. Let there by angles *PQ* and *RM* at the center of both lines, bisecting a semi-circle into two equal length arcs. In a circle, if both lines are drawn from the point *s*, will always be equal in length and the values of the angles on both sides and both sections of the two chords bisecting the circle into two arcs, will be equal.

QED:

Here we have the exchange of virtual photons. Each electrically charge particle, i.e. proton and electron, emit and absorb streams of virtual photons according to the inverse square law along a two-dimensional phase space in a three-dimensional 360-degree cone.

Proposition XXXXV. Theorem VIII

Given an orbiting body, if the body is a satellite or moon, it will revolve around the larger orbiting body by a different G.R.F.F.. *This behaves according to the derived mechanical laws similar to the general* G.R.F.F..

If we were to draw a line through the center of the second larger body in question, emanating from the center of the largest body in question, we derive the following principles. We will have two separate G.R.F.F.'s rather than the singular G.R.F.F.. Now when the smallest orbiting body in question, in this case the moon or satellite, the distance from the second larger orbiting body, will be the same on the both sides drawn directly straight through larger bodies and generally closer than when it is on the side of the second orbiting body in question, where it is directly perpendicular to the line drawn through the two larger bodies in question. Then lines *AB* and *BC* will be equal and always larger to the lines *DE* and *EF*, which are also equal to each other.

The equations of motion of the moon take two parts, two different covariant tensors that represent both lunar gravitational fields. If we were to draw a line from the center of the largest body in question, *S* through an orbiting smaller body *T* when the orbiting smallest body *P* is positioned directly parallel to that line in which it crosses through the center of P we employ the following equation of motion

$$\left\{ (\sin\phi du \, / \, dt) + \sum_{i=1}(\sin\varphi dv \, / \, dt + \cos\varphi dv \, / \, dt) + \sum_{j=1}(\sin\varphi dv \, / \, dt + \cos\varphi dv \, / \, dt) \right\} \frac{M_\odot m_1 m_2}{r^2} \quad (1.388)$$

and from a position directly perpendicular or at a right angle to that original position our second field tensor will take the form

$$\left\{ \sin\alpha dv \, / \, dt + \sum_{j=1}(\sin\varphi dv \, / \, dt + \cos\varphi dv \, / \, dt) \right\} \frac{Mm}{r^2}$$

$$+\left\{ (-\sin\alpha dv \, / \, dt) + \sum_{i=1}(\sin\varphi dv \, / \, dt + \cos\varphi dv \, / \, dt) \right\} \frac{Mm}{r^2} \quad (1.389)$$

$$+\left\{ (\sin\phi du \, / \, dt) + \sum_{i=1}(\sin\varphi dv \, / \, dt + \cos\varphi dv \, / \, dt) + \sum_{j=1}(\sin\varphi dv \, / \, dt + \cos\varphi dv \, / \, dt) \right\} \frac{m_1 m_2}{r^2}$$

The disturbing function

The equations of motion for three bodies under the same mutual attraction will be given as

$$m_1 \frac{d^2 r_1}{dt^2} = G \frac{m_1 m_2}{r_{12}^3}(r_2 - r_1) + G \frac{m_1 m_3}{r_{13}^3}(r_3 - r_1)$$

$$m_2 \frac{d^2 r_2}{dt^2} = G \frac{m_2 m_1}{r_{21}^3}(r_1 - r_2) + G \frac{m_2 m_3}{r_{23}^3}(r_3 - r_2) \quad (1.390)$$

$$m_3 \frac{d^2 r_3}{dt^2} = G \frac{m_3 m_1}{r_{31}^3}(r_1 - r_3) + G \frac{m_3 m_2}{r_{32}^3}(r_2 - r_3)$$

The equation representing the motion of the moon, the lunar or three-body problem, can be given as such in four separate parts, unified into one singular equation

$$m\frac{d^2r_1}{dt} = \begin{pmatrix} G_1\frac{M\ m_1}{r_{12}^2}\big((R)-(r)\big)\sin\varphi + G_2\frac{M\ m_1}{r_{12}^2}\big((R)-(r)\big)(\cos+\sin)\varphi_i \\[2mm] +G_3\frac{M\ m_1}{r_{12}^2}\big((R)-(r)\big)(\cos+\sin)\varphi_j + G_4\frac{M\ m_1}{r_{12}^2}\big((R)-(r)\big)(\cos+\sin)\varphi_k \\[2mm] +G_5\frac{M\ m_1}{r_{12}^2}\big((R)-(r)\big)(\cos+\sin)\varphi_l \end{pmatrix}$$

$$+\begin{pmatrix} -G_6\frac{m_2m_1}{r_{12}^2}(R)\sin\varphi + (-)G_7\frac{m_2m_1}{r_{12}^2}(R)(\cos+\sin)\varphi_i \\[2mm] +(-)G_8\frac{m_2m_1}{r_{12}^2}(R)(\cos+\sin)\varphi_j + (-)G_9\frac{m_2m_1}{r_{12}^2}(R)(\cos+\sin)\varphi_k \\[2mm] +(-)G_{10}\frac{m_2m_1}{r_{12}^2}(R)(\cos+\sin)\varphi_l \end{pmatrix}$$

(1.391)

and

$$m\frac{d^2r_2}{dt} = \begin{pmatrix} (-)G_1\frac{M\ m_1}{r_{12}^2}\big((R)+(r)\big)\sin\varphi + (-)G_2\frac{M\ m_1}{r_{12}^2}\big((R)+(r)\big)(\cos+\sin)\varphi_i \\[2mm] +(-)G_3\frac{M\ m_1}{r_{12}^2}\big((R)+(r)\big)(\cos+\sin)\varphi_j \\[2mm] +(-)G_4\frac{M\ m_1}{r_{12}^2}\big((R)+(r)\big)(\cos+\sin)\varphi_k \\[2mm] +(-)G_5\frac{M\ m_1}{r_{12}^2}\big((R)+(r)\big)(\cos+\sin)\varphi_l \end{pmatrix}$$

(1.392)

$$+\begin{pmatrix} G_6\frac{m_2m_1}{r_{12}^2}(R)\sin\varphi + G_7\frac{m_2m_1}{r_{12}^2}(R)(\cos+\sin)\varphi_i \\[2mm] +G_8\frac{m_2m_1}{r_{12}^2}(R)(\cos+\sin)\varphi_j + G_9\frac{m_2m_1}{r_{12}^2}(R)(\cos+\sin)\varphi_k \\[2mm] +G_{10}\frac{m_2m_1}{r_{12}^2}(R)(\cos+\sin)\varphi_l \end{pmatrix}$$

then

$$m\frac{d^2r_3}{dt^2} = \left(G_1 \frac{M\ m_1}{r_{12}^2}(R)(\cos+\sin)\varphi_i + G_2 \frac{M\ m_1}{r_{12}^2}(R)(\cos+\sin)\varphi_j \right)$$
$$+ \left((-)G_3 \frac{M\ m_1}{r_{12}^2}(R)(\cos+\sin)\varphi_i + (-)G_4 \frac{M\ m_1}{r_{12}^2}(R)(\cos+\sin)\varphi_j \right)$$

(1.393)

and then we will have

$$m\frac{d^2r_3}{dt^2} = \left((-)G_1 \frac{M\ m_1}{r_{12}^2}(R)(\cos+\sin)\varphi_j + (-)G_2 \frac{M\ m_1}{r_{12}^2}(R)(\cos+\sin)\varphi_i \right)$$
$$+ \left(G_3 \frac{M\ m_1}{r_{12}^2}(R)(\cos+\sin)\varphi_j + G_4 \frac{M\ m_1}{r_{12}^2}(R)(\cos+\sin)\varphi_i \right)$$

(1.394)

For the equations representing the Earth-Moon relations we will have

$$G_E \frac{m_2 m_1}{r^2} = \begin{pmatrix} G_6 \frac{m_2 m_1}{r_{12}^2}(R)\sin\varphi + G_7 \frac{m_2 m_1}{r_{12}^2}(R)(\cos+\sin)\varphi_i \\ +G_8 \frac{m_2 m_1}{r_{12}^2}(R)(\cos+\sin)\varphi_j + G_9 \frac{m_2 m_1}{r_{12}^2}(R)(\cos+\sin)\varphi_k \\ +G_{10} \frac{m_2 m_1}{r_{12}^2}(R)(\cos+\sin)\varphi_l \end{pmatrix}$$
$$+ \begin{pmatrix} (-)G_6 \frac{m_2 m_1}{r_{12}^2}(R)\sin\varphi + (-)G_7 \frac{m_2 m_1}{r_{12}^2}(R)(\cos+\sin)\varphi_i \\ +(-)G_8 \frac{m_2 m_1}{r_{12}^2}(R)(\cos+\sin)\varphi_j + (-)G_9 \frac{m_2 m_1}{r_{12}^2}(R)(\cos+\sin)\varphi_k \\ +(-)G_{10} \frac{m_2 m_1}{r_{12}^2}(R)(\cos+\sin)\varphi_l \end{pmatrix}$$

(1.395)

The components of $F = (F_r, F_\alpha, F_h)$

We focus here on the relative direction, R, of the Sun on the ecliptic plane, relative to the Earth. Specifically

$$\angle v = \angle(v,r) \to \angle\omega : \angle\psi = \angle(R,r) : \angle U = \angle(R,\omega)$$

(1.396)

then for R we will have

$$R = R\left[i\cos\left(U + \Omega\right) + j\sin\left(U + \Omega\right)\right] = R\left(+\cos U \cos \omega + \sin U \sin \omega \cos \iota\right)$$

$$R_1 = \left\{R_1\left[i\cos\left(U + \Omega\right) + j\sin\left(U + \Omega\right)\right]\right\}\alpha = \left\{R\left(+\cos U \cos \omega + \sin U \sin \omega \cos \iota\right)\right\}\alpha \quad (1.397)$$

$$R_2 = \left\{R_2\left[i\cos\left(U + \Omega\right) + j\sin\left(U + \Omega\right)\right]\right\}\beta = \left\{R\left(+\cos U \cos \omega + \sin U \sin \omega \cos \iota\right)\right\}\beta$$

Then following this

$$F_T = \pm N_\alpha \left(F_S \oplus F_E\right)r\left(t\right) = \pm N_\alpha \left(F_S \oplus F_E\right)\left(x\left(t\right)e_x + y\left(t\right)e_y + z\left(t\right)e_z\right) \qquad (1.398)$$

and

$$F_T = \pm N_\alpha \left(F_S \oplus F_E\right)\left(\beta\right)r\left(t\right) = \pm N_\alpha \left(F_S \oplus F_E\right)\left(\beta\right)\left(x\left(t\right)e_x + y\left(t\right)e_y + z\left(t\right)e_z\right) \quad (1.399)$$

and finally

$$F_T = \pm N_\alpha \left(F_S \oplus F_E\right)\left(\gamma\right)r\left(t\right) = \pm N_\alpha \left(F_S \oplus F_E\right)\left(\gamma\right)\left(x\left(t\right)e_x + y\left(t\right)e_y + z\left(t\right)e_z\right) \quad (1.400)$$

as long as

$$\cos\psi = \cos U \cos\omega + \sin U \sin\omega\cos\tau = \cos\angle\left(R, r\right) \qquad (1.401)$$

Application pertaining to the lunar motion

For the Earth-Moon system we will have

$$F_E \rightarrow Earth - Moon \qquad (1.402)$$

With these definitions and with the expressions for the components of $F = \left(F_i, F_{jk}, F_{mn}\right)$ given in the previous equations, the variational equations as applied to the motion of the Moon, gives us

$$\frac{1}{h}\frac{dh}{dt} = \pm N_\alpha \left(F_S \oplus F_E\right)\frac{r^2}{a^2\sqrt{\left(1-e^2\right)}}r\left(t\right)$$

$$= \pm N_\alpha \left(F_S \oplus F_E\right)\frac{r^2}{a^2\sqrt{\left(1-e^2\right)}}\left(x\left(t\right)e_x + y\left(t\right)e_y + z\left(t\right)e_z\right)$$

$$(1.403)$$

211

where

$$R_1 = \pm R\left(\cos U \cos \omega + \sin U \sin \omega \cos \iota\right)$$

$$R_2 = R\left(\cos U \cos \omega + \sin U \sin \omega \cos \iota\right)$$

$$R_3 = \pm \left\{ \begin{array}{l} R\left(\cos U \cos \omega + \sin U \sin \omega \cos \iota\right) \\ +\left(\cos R\left(\cos U \cos \omega + \sin U \sin \omega \cos \iota\right) + \sin R\left(\cos U \cos \omega + \sin U \sin \omega \cos \iota\right)\right) \end{array} \right\}$$

$$R_4 = \left\{ \begin{array}{l} R\left(\cos U \cos \omega + \sin U \sin \omega \cos \iota\right) \\ +\left(\cos R\left(\cos U \cos \omega + \sin U \sin \omega \cos \iota\right) + \sin R\left(\cos U \cos \omega + \sin U \sin \omega \cos \iota\right)\right) \end{array} \right\} \quad (1.404)$$

$$R_5 = \pm \left\{ \begin{array}{l} R\left(\cos U \cos \omega + \sin U \sin \omega \cos \iota\right) \\ +\left(\cos R\left(\cos U \cos \omega + \sin U \sin \omega \cos \iota\right) - \sin R\left(\cos U \cos \omega + \sin U \sin \omega \cos \iota\right)\right) \end{array} \right\}$$

$$R_6 = \left\{ \begin{array}{l} R\left(\cos U \cos \omega + \sin U \sin \omega \cos \iota\right) \\ +\left(\cos R\left(\cos U \cos \omega + \sin U \sin \omega \cos \iota\right) - \sin R\left(\cos U \cos \omega + \sin U \sin \omega \cos \iota\right)\right) \end{array} \right\}$$

then

$$\frac{d\iota}{dt} = \pm N_\beta \left(F_S \oplus F_E\right) \frac{r^2}{a^2 \sqrt{\left(1-e^2\right)}} r(t)$$

$$= \pm N_\beta \left(F_S \oplus F_E\right) \frac{r^2}{a^2 \sqrt{\left(1-e^2\right)}} \left(x(t)e_x + y(t)e_y + z(t)e_z\right) \quad (1.405)$$

$$= \pm N_\beta \left(r \cos v\right)\left(F_S \oplus F_E\right) r(t)$$

$$= \pm N_\beta \left(r \cos v\right)\left(F_S \oplus F_E\right)\left(x(t)e_x + y(t)e_y + z(t)e_z\right)$$

then we will have

$$\frac{de}{dt} = \pm N_\beta \left(F_S \oplus F_E\right) \frac{r\sqrt{\left(1-e^2\right)}}{a} r$$

$$= \pm N_\beta \left(F_S \oplus F_E\right) \frac{r\sqrt{\left(1-e^2\right)}}{a} \left(x(t)e_x + y(t)e_y + z(t)e_z\right) \quad (1.406)$$

and

$$\frac{d\omega}{dt} = \pm N_\beta \left(F_S \oplus F_E\right) \frac{\sqrt{\left(1-e^2\right)}}{e} \frac{r}{a} \cos \varphi r$$

$$= \pm N_\beta \left(F_S \oplus F_E\right) \frac{r\sqrt{\left(1-e^2\right)}}{a} \frac{r}{a} \cos \varphi \left(x(t)e_x + y(t)e_y + z(t)e_z\right)$$

(1.407)

Ignoring the first term in F_r (with the factor e) in the equation for da/dt and ignoring the factor $\left(1-e^2/r^2\right)^{1/2}$ we will have

$$\frac{da}{dt} \pm N_\beta \left(F_S \oplus F_E\right) \frac{a}{1} r = \pm N_\beta \left(F_S \oplus F_E\right) \frac{a}{1}\left(x(t)e_x + y(t)e_y + z(t)e_z\right)$$

(1.408)

then since $\frac{r}{R}$ where r=radius of Moon's orbit and R=radius of Earth's orbit.

13

The three-body problem: Lunar theory part 2

Introduction

This the second part of the newly formed theory of the three-body problem.

Proposition *XXXXVI*. Theorem *I*

Bodies, whose forces decrease as the square of the distance from their centers, could move among themselves in ellipses; and also by the radii drawn to the focus could describe areas very nearly proportional to the times.

Case. *I*. Given a larger body, denoted as M, about which several smaller bodies, are revolving. The center of gravity of all of these masses will either be at rest or moving uniformly in a right line. If the smaller bodies are small enough, then the center of gravity will not be very different from the location of M, which is either at rest or moving in a right line; about which the smaller bodies will revolve. If the mutual gravitational behavior between the smaller bodies are tiny as compared to the gravitational force of M, then the smaller bodies will each revolve about M in ellipses describing equal areas in time.

Case *II*. We consider a similar case as to case *I*. The difference is that the gravitational behavior between bodies or masses, say m and m, is sufficiently stronger that they describe elliptic orbits about their common center of gravity. If the larger mass *M*, is sufficiently farther away from that of m and m, which is small compared to the distance *MG*, *M* will affect the motion of the two smaller masses, m and m, such that they are compelled by parallel and equal accelerations. Since equal and parallel accelerations will not affect their relative motions it follows then that M will act on m and m as if they were a singular mass or body. Then the smaller masses will follow a parabola or hyperbola about M when the nature of the force is less and will follow the path of an ellipse when the force is greater.

Cor. I. If parts of the previously defined system move in ellipses without any perturbation, it follows then that they are affected, motion wise, by the presence of other bodies, this force is weak or else is impressed very nearly equally and in parallel directions upon all of them.

Proposition *XXXXVII*. Theorem *II*

Given three bodies, whose G.R.F.F. *decreases as the square of the distances, affect each other according to the laws of the* G.R.F.F.*'s; and the* G.R.F.F. *influences of any two bodies on a third be between themselves inversely as the squares of the distances; and the two smaller bodies revolve about the larger body; the interior of the two revolving bodies will, by radii drawn to the innermost and largest, describe round that body areas proportional to the times, and a figure approaching that of un ellipse having its focus in the point of intersection of the radii, the larger body affected slightly by the presence of the smaller bodies, than as if the larger body were not affected by the smaller bodies, but remained at rest.*

Case I. Given three bodies (T, P and S) then the perturbation of the elliptic orbit, described by the smaller body P about the larger body T, is by another smaller body S. S is a lesser body because in the sense that its influence on the motion of P may be considered as a smaller perturbation of the unperturbed orbit in which P describes about T. Here S is considered the Sun, T the Earth, and P the moon.

Here the orbits of P about T and T about S are considered coplanar. When undisturbed, P, will describe an elliptic orbit about T, as its focus. Here the presence and influence of S does not actually follow the principles of a disturbing force but rather a key part of the original elliptic orbit of P about T.

Let TS represent the line force in which constitutes the central direction of the singular force vector in which we find the application of part of the overall *G.R.F.F.,* or partial force tensor. This is the representation of the partial force tensor, TS, as exerted on T by the presence of S.

Here LS represents a partial force tensor on P by the presence of S. If LS is placed along LM

parallel to PT and MS we will have

$$LS = LM + MS \qquad (1.409)$$

Of these forces, LM being parallel to PT, we find the direction of the force tensors acting on P at that singular point. The central character of this force, the force tensor on P, will preserve the law of areas. The force MS, no longer parallel to PT, will still allow for the preservation of the law of areas and the keeping the original structure of the ellipse, regardless of the position of P about T relative to S. The value of the eccentricity is directly proportional to the inverse of the square of the distance from S. Since there are equal accelerations imparted to T and P by S, It will not affect their relative motion, so that we will have the following relation

$$TS - LS = TL = TM - LM \qquad (1.410)$$

Both TM and TS will affect the relative motion of P and T. Then of the two forces associated with LM and TM that compute directly in the motion of P about S, the law of areas is preserved as is the original structure of the ellipse.

Case *II.* We next consider the case of how the non-coplanarity of the orbits P and T will affect the results arrived at case I. The expression for F_h, in the direction of the angular momentum and normal to the orbital plane, given earlier

$$F_T = \left(F_S \oplus F_E\right)\theta_i = \left(F_S \oplus F_E\right)\left(\sin\theta + \cos\theta\right)_i \qquad (1.411)$$

where ι is the angle of inclination and U the angle between the line of the nodes and direction of TS.

.

Proposition XXXXVIII: Corollaries I-VI

In these corollaries there is no distinction between ordinary elliptical orbits and perturbed orbits and the orbit is also coplanar with the orbit of T.

The perturbing function

In the above figure the orbit described by P is circular until we introduce the additional forces added by S. An additional line TK is drawn perpendicular to SL. Then given $ST \quad PK$ we will have

$$ST \cong SK \qquad (1.412)$$

We consider the following geometrical relations

$$LS - TS = LT \qquad (1.413)$$

Resolve LT into its orthogonal components

$$LT = LE + ET \qquad (1.414)$$

both perpendicular and parallel to TP. Then by equation (1.412) we will have

$$ST \quad SP + PK : SL = SP + PL \qquad (1.415)$$

then we will have

$$LS = ST^3 / SP^2 \qquad (1.416)$$

therefore

$$3SP^2 \cdot PK + 3SP \cdot PK^2 + PK^3 = (SP + PK) - SP^3 \quad ST^3 - SP^3$$
$$= LS \cdot SP^2 - SP^3 = SP^2(LS - SP) = SP^2 \cdot PL \tag{1.417}$$

Then by dividing by SP^2

$$3PK\left(1 + \frac{PK}{SP} + \frac{1}{3}\frac{PK^2}{SP^2}\right) \quad PL \tag{1.418}$$

since $PK \quad SP$ we will have

$$3PK = PL + O(PT/ST) \tag{1.419}$$

The triangle EPL being right-angled

$$LE = PL\sin\angle EPL \quad 3PK\sin\angle TPK$$
$$TE + TP = PL\cos\angle TPK \quad 3PK\cos\angle TPK \tag{1.420}$$

The following forces are measured according to the unit $R = ST$, the forces are then given as

$$F_r = \left\{\left(\sin G_1 \frac{M \ m_2}{r^3} R + (-\sin)G_2 \frac{M \ m_2}{r^3} R\right)\right\}\sin\phi \tag{1.421}$$

And

$$F_T = (F_s \oplus F_E)\theta_i = (F_s \oplus F_E)(\sin\theta + \cos\theta)_i \tag{1.422}$$

The G.R.F.F.

The ordinary motion of the moon without the contribution of the Sun can be given as

$$K = \begin{pmatrix} \left| (-)\sin G_1 \dfrac{m_2 m_1}{r_{32}^2}(r_2 - r_1) + \left((-)G_2 \dfrac{m_2 m_1}{r_{32}^2}(r_3 - r_1) \right)(\sin + \cos)_i \right. \\ \\ \left. + \left((-)G_3 \dfrac{m_2 m_1}{r_{32}^2}(r_3 - r_1) \right)(\sin + \cos)_j \right. \\ \\ \left. + \left((-)G_5 \dfrac{m_2 m_1}{r_{32}^2}(r_3 - r_1) \right)(\sin + \cos)_k + (-)G_6 \dfrac{m_2 m_1}{r_{32}^2}(r_3 - r_2)(\sin + \cos)_l \right| \end{pmatrix}$$

$$+ \begin{pmatrix} \left| \sin G_7 \dfrac{m_2 m_1}{r_{32}^2}(r_2 - r_1) + \left(G_8 \dfrac{m_2 m_1}{r_{32}^2}(r_3 - r_1) \right)(\sin + \cos)_i \right. \\ \\ \left. + \left(G_9 \dfrac{m_2 m_1}{r_{32}^2}(r_3 - r_1) \right)(\sin + \cos)_j \right. \\ \\ \left. + \left(G_{10} \dfrac{m_2 m_1}{r_{32}^2}(r_3 - r_1) \right)(\sin + \cos)_k + G_{11} \dfrac{m_2 m_1}{r_{32}^2}(r_3 - r_2)(\sin + \cos)_l \right| \end{pmatrix}$$

(1.423)

but since the presence of the Sun is a key factor in the lunar motion of the moon as a whole will then be given as

$$F_S = \begin{pmatrix} \left(\sin G_1 \dfrac{M_2 m_1}{r_{32}^2}(R-\rho) + \left(G_2 \dfrac{M_2 m_1}{r_{32}^2}(R-\rho) \right)(\sin+\cos)_i \right) \\[2ex] + \left(G_3 \dfrac{M_2 m_1}{r_{32}^2}(R-\rho) \right)(\sin+\cos)_j \\[2ex] + \left(G_5 \dfrac{M_2 m_1}{r_{32}^2}(R-\rho) \right)(\sin+\cos)_k + G_6 \dfrac{M_2 m_1}{r_{32}^2}(R-\rho)(\sin+\cos)_l \end{pmatrix}$$

$$+ \begin{pmatrix} \left(\left(G_6 \dfrac{M_2 m_1}{r_{32}^2}(R-\rho) \right)(\sin+\cos)_i \left(G_7 \dfrac{M_2 m_1}{r_{32}^2}(R-\rho) \right)(\sin+\cos)_k \right) \\ + G_8 \dfrac{M_2 m_1}{r_{32}^2}(R-\rho)(\sin+\cos)_l \\[2ex] + \left(\left(G_9 \dfrac{M_2 m_1}{r_{32}^2}(R-\rho) \right)(\sin+\cos)_j \left(G_{10} \dfrac{M_2 m_1}{r_{32}^2}(R-\rho) \right)(\sin+\cos)_m \right) \\ + G_{11} \dfrac{M_2 m_1}{r_{32}^2}(R-\rho)(\sin+\cos)_n \end{pmatrix}$$

$$\begin{pmatrix} \left((-)\sin G_1 \dfrac{M_2 m_1}{r_{32}^2}(R+\rho) + \left((-)G_2 \dfrac{M_2 m_1}{r_{32}^2}(R+\rho) \right)(\sin+\cos)_i \right) \\[2ex] + \left((-)G_3 \dfrac{M_2 m_1}{r_{32}^2}(R+\rho) \right)(\sin+\cos)_j \\[2ex] + \left((-)G_5 \dfrac{M_2 m_1}{r_{32}^2}(R+\rho) \right)(\sin+\cos)_k + (-)G_6 \dfrac{M_2 m_1}{r_{32}^2}(R+\rho)(\sin+\cos)_l \end{pmatrix}$$

$$+ \begin{pmatrix} \left(\left((-)G_6 \dfrac{M_2 m_1}{r_{32}^2}(R+\rho) \right)(\sin+\cos)_i \left((-)G_7 \dfrac{M_2 m_1}{r_{32}^2}(R+\rho) \right)(\sin+\cos)_k \right) \\ + (-)G_8 \dfrac{M_2 m_1}{r_{32}^2}(R+\rho)(\sin+\cos)_l \\[2ex] + \left(\left((-)G_9 \dfrac{M_2 m_1}{r_{32}^2}(R+\rho) \right)(\sin+\cos)_j \left((-)G_{10} \dfrac{M_2 m_1}{r_{32}^2}(R+\rho) \right)(\sin+\cos)_m \right) \\ + (-)G_{11} \dfrac{M_2 m_1}{r_{32}^2}(R+\rho)(\sin+\cos)_n \end{pmatrix} \tag{1.424}$$

in terms of the angular momentum, that is the conservation of the angular momentum, we will have

$$K + F_S = \left\{ G_2 \frac{m_2 m_1}{r_{32}^2} \left((r_3 - r_1) - (r_2 - r_3) \right) + G_2 \frac{m_2 m_1}{r_{32}^2} \left((R) - (\rho) \right) + G_2 \frac{m_2 m_1}{r_{32}^2} \left((R) - (\rho) \right) \right\}$$

$$+ \left\{ \begin{array}{c} \left\{ \sin G_1 \frac{M \ m_1}{r^3} R + (-\sin) G_2 \frac{M \ m_1}{r^3} R \right\} \sin \phi \\[2mm] + \left\{ + \left(G_3 \frac{M \ m_1}{r^3} R + G_4 \frac{M \ m_1}{r^3} R \right) (\sin \varphi + \cos \varphi)_i \\[2mm] + \left(G_5 \frac{M \ m_1}{r^3} R + G_6 \frac{M \ m_1}{r^3} R \right) (\sin \varphi + \cos \varphi)_j \right\} \end{array} \right\} \tag{1.425}$$

and for the constant alpha we will have

$$K + F_h = \left\{ G_2 \frac{m_2 m_1}{r_{32}^2} \left((r_3 - r_1) - (r_2 - r_3) \right) + G_2 \frac{m_2 m_1}{r_{32}^2} \left((R) - (\rho) \right) + G_2 \frac{m_2 m_1}{(r_{21})_1^2} \left((R) + (\rho) \right) \right\}$$

$$+ \left\{ \begin{array}{c} \left\{ G \frac{M \ m_1}{r^3} R + (-) G \frac{M \ m_1}{r^3} R + G \frac{M \ m_1}{(r)_1^3} (R + (r)) \right\} \times m_1 (R) \\[2mm] + \left\{ (-) G \frac{M \ m_1}{r^3} R + G \frac{M \ m_1}{r^3} R + G \frac{M \ m_1}{(r)_1^3} (R - (r)) \right\} \times m_1 (R) \end{array} \right\}$$

$$= \left\{ G_2 \frac{m_2 m_1}{r_{32}^2} \left((r_3 - r_1) - (r_2 - r_3) \right) + G_2 \frac{m_2 m_1}{r_{32}^2} \left((R) - (\rho) \right) + G_2 \frac{m_2 m_1}{(r_{21})_1^2} \left((R) + (\rho) \right) \right\}$$

$$+ \left\{ \begin{array}{c} \left\{ G \frac{M \ m_1}{r^3} R + (-) G \frac{M \ m_1}{r^3} R + G \frac{M \ m_1}{(r)_1^3} (R + (r)) \right\} \\[2mm] + \left\{ (-) G \frac{M \ m_1}{r^3} R + G \frac{M \ m_1}{r^3} R + G \frac{M \ m_1}{(r)_1^3} (R - (r)) \right\} \end{array} \right\} \times m_1 (R) \tag{1.426}$$

Then

$$F_T = F_M \frac{m_E m_M}{r^3} + F_S \frac{M \ m_M}{r^3} \tag{1.427}$$

Measuring r in the units of the ordinary radius orbit of the elliptical orbit of *P*, its values at syzygies and at quadratures be

$$r_1 = 1 - x : r_0 1 + x \tag{1.428}$$

The forces F and $F1$ at syzygies and at quadratures are

$$F_1 = \frac{K}{(1-x)^2}\left(1 \pm \frac{1}{K}(P_M)\right) : F_0 = \frac{K}{(1+x)^2}\left(1 \pm 2\frac{1}{K}(P_M)\right)$$

(1.429)

The Suns contribution as a necessary force in the elliptical structure

Here the orbit is an prolate ellipse centered at T; and its minor and major axes, in the direction of the syzygies

$$r = 1 - x\cos 2\psi \rightarrow 1 + x\cos\left[2(1 \pm (F_T))\omega\right] = 1 + x\cos\left[2(1 \pm (F_T))\omega\right]d\omega$$

(1.430)

then written in terms of components according to the inverse-square law we will have

$$r_1 = \frac{1 + x\cos\left[2(1 \pm (F_T))\omega\right]}{r^3}(R-r) = \frac{1 + x\cos\left[2(1 \pm (F_T))\omega\right]}{r^3}(R-r)$$

$$r_2 = \frac{1 + x\cos\left[2(1 \pm (F_T))\omega\right]}{r^3}(R+r) = \frac{1 + x\cos\left[2(1 \pm (F_T))\omega\right]}{r^3}(R+r)$$

(1.431)

rewritten in terms of r the radius of curvature will be given as

$$\rho = \frac{\left[r^2 + (dr/d\psi)^2\right]^{3/2}}{r^2 + 2(dr/d\psi)^2 - rd^2r/d\psi^2} \quad \frac{r^2}{r - d^2r/d\psi^2}$$

(1.432)

then for r given in equation (1.430)

$$\rho = \frac{1 + 2\dfrac{x\cos 2(1 \pm (F_T))\omega}{r^3}(R-r)}{1 + \left[1 + 4(1 \pm (F_T))^2\right]\dfrac{x\cos 2(1 \pm (F_T))\omega}{r^3}(R-r)}$$

$$= \frac{1 + 2\dfrac{x\cos 2(1 \pm (F_T))\omega}{r^3}(R-r)}{1 + \left[1 + 4(1 \pm (F_T))^2\right]\dfrac{x\cos 2(1 \pm (F_T))\omega}{r^3}(R-r)}$$

(1.433)

For r we will have then

$$\frac{\rho_0}{\rho_1}\frac{1+2x}{1-2x}\frac{1-x\left[1+4\left(1\pm\left(F_T\right)\right)^2\right]}{1+x\left[1+4\left(1\pm\left(F_T\right)\right)^2\right]}=\frac{1+2x}{1-2x}\frac{1-x\left[1+4\left(1\pm\left(F_T\right)\right)^2\right]}{1+x\left[1+4\left(1\pm\left(F_T\right)\right)^2\right]}$$

$$1+2x\left[4\left(1\pm\left(F_T\right)\right)^2-1\right]=1+2x\left[4\left(1\pm\left(F_T\right)\right)^2-1\right]$$

(1.434)

The variation of the 'constant of areas'

To determine the constant of areas we use previously developed formulas starting with

$$\frac{dh}{dt}=F_T=\left(\pm N_\alpha r^2\left(F_S\oplus F_E\right)\right)m_1\cdot r=\left(\pm N_\alpha r^2\left(F_S\oplus F_E\right)\right)m_1\cdot\left(x_y e_1+x_y e_2+x_z e_3\right)$$ (1.435)

then

$$dh=\left(\pm N_\alpha r^2\left(F_S\oplus F_E\right)\theta_i\right)\cdot m_1\cdot r\left(dt\right)=\left(\pm N_\alpha r^2\left(F_S\oplus F_E\right)\right)m_1\cdot\left(x_y e_1+x_y e_2+x_z e_3\right)dt$$
$$=\left(\pm N_\alpha r^2\left(F_S\oplus F_E\right)\theta_i\right)\cdot m_1\cdot dv=\left(\pm N_\alpha r^2\left(F_S\oplus F_E\right)\left(\sin\theta+\cos\theta\right)_i\right)m_1\cdot dv$$

(1.436)

for an ordinary orbit we will have

$$ndt=d\omega$$

(1.437)

we will then have

$$dh=\left(\pm N_\alpha r^2\left(F_S\oplus F_E\right)\right)\cdot m_1\cdot rd\omega=\left(\pm N_\alpha r^2\left(F_S\oplus F_E\right)\right)\cdot m_1\cdot\left(x_y e_1+x_y e_2+x_z e_3\right)d\omega$$ (1.438)

we will then have

$$h=\left(\pm N_\alpha r^2\left(F_S\oplus F_E\right)r\right)\times m_1=\left(\pm N_\alpha r^2\left(F_S\oplus F_E\right)\left(x_y e_1+x_y e_2+x_z e_3\right)\right)\times m_1$$

(1.439)

The determination of x

We start with

$$V^2 = \left(\frac{dr}{dt}\right)^2 + (h)^2 \tag{1.440}$$

The force given in terms of this is

$$F = \frac{V^2}{\rho} \tag{1.441}$$

where ρ denotes the radius of curvature. At syzygies and at quadratures, the above equation can be given as

$$\frac{F_1}{F_0} = \frac{V_1^2 \rho_0}{V_0^1 \rho_1} = \frac{\dfrac{K}{(1-x)^2}\left(1 \pm \dfrac{1}{K}(F_T)\right)}{\dfrac{K}{(1+x)^2}\left(1 \pm \dfrac{1}{K}(F_T)\right)} \tag{1.442}$$

Cor. I. If there are several bodies P, S, R, and such, revolve about a greater body T, the motion of the innermost revolving body P will be the least affected by the presence of the other bodies according to the law of the G.R.F.F..

The contribution of the generalized force F_T in relation to F_{1T} and F_{2T} acting on the smaller bodies, is directly proportional to the radius of its orbit about T.

Cor. II. In a system of three bodies T, P and S, the gravitational behavior between two bodies in relation to a third larger body be to each other inversely as the squares of the distances, the body P, by the radius PT, will describe its area about the body T quicker near the conjunction A and the Opposition B. Then it will near the quadratures C and D.

Cor. III. From the same reasoning it seems that the body P, other things being the same, more quicker in the conjunction and opposition that in the quadratures.

If follows then that

$$V_1 > V_0 \qquad (1.443)$$

Cor. IV. The orbit of the body P, other things being the same, is more curved at the quadratures then at the conjunction and opposition.

Then we will have

$$\rho_0^{-1} > \rho_1^{-1} \qquad (1.444)$$

Cor. V. The body P goes, other things being the same, goes farther from the body T at the quadratures than at the conjunction and opposition.

Cor. VI. The periodical time would be increased and diminished in a ratio compounded of the $3/2$ th power of the ratio of the radius, and of the square root of that ratio in which the $G.R.F.F.$ related to the central body T was diminished or increased, by the increase or decrease of the action related to the distant body S.

We begin with

$$\frac{da}{dt} = \frac{2a/r^2}{r} \cdot F_T = \frac{2a/r^2}{r} \cdot \left(\pm N_\alpha r^2 \left(F_S \oplus F_E \right) r_i \right)$$
$$= \frac{2a/r^2}{r} \cdot \left(\pm N_\alpha r^2 \left(F_S \oplus F_E \right) \left(x_y e_1 + x_y e_2 + x_z e_3 \right)_i \right) \qquad (1.445)$$

we will then have

$$\frac{(G.R.F.F.)_s}{(G.R.F.F.)_q} = \frac{r_s}{r_q} \left(\frac{P_q}{P_s} \right)^2 \qquad (1.446)$$

where the subscripts s and q refer to syzygies and quadratures. It is noted that we have not distinguished between r_s and r_q when r occurs with the small factor η^2. P_q and P_s denote the quadratures and syzygies. Then

$$P \propto r^{3/2} \left(\pm N_\alpha r^2 \left(F_S \oplus F_E \right) r \right) = r^{3/2} \left(\pm N_\alpha r^2 \left(F_S \oplus F_E \right) \left(x_y e_1 + x_y e_2 + x_z e_3 \right) \right) \qquad (1.447)$$

then for n we will have

$$n = \left(\frac{1}{a^3}\right)^{1/2}\left(\sin G_1 \frac{m_2 m_1}{r_{21}^2}(r_2 - r_1) + (-\sin)G_2 \frac{m_2 m_1}{r_{21}^2}(r_2 - r_1)\right)$$

$$= \left(\frac{1}{a^3}\right)^{1/2}\left(\sin G_1 \frac{m_2 m_1}{r_{21}^2}((r_3 - r_1) + (r_2 - r_3)) + (-\sin)G_1 \frac{m_2 m_1}{r_{21}^2}((r_3 - r_1) + (r_2 - r_3))\right) \quad (1.448)$$

$$= \left(\frac{1}{a^3}\right)^{1/2}\left(\sin G_1 \frac{m_2 m_1}{r_{21}^2}((R) + (\rho)) + (-\sin)G_1 \frac{m_2 m_1}{r_{21}^2}((R) + (\rho))\right)$$

and for a we will have

$$a = \left(\frac{1}{n^2}\right)^{1/3}\left(\sin G_1 \frac{m_2 m_1}{r_{21}^2}(r_2 - r_1) + (-\sin)G_2 \frac{m_2 m_1}{r_{21}^2}(r_2 - r_1)\right)$$

$$= \left(\frac{1}{n^2}\right)^{1/3}\left(\sin G_1 \frac{m_2 m_1}{r_{21}^2}((r_3 - r_1) + (r_2 - r_3)) + (-\sin)G_1 \frac{m_2 m_1}{r_{21}^2}((r_3 - r_1) + (r_2 - r_3))\right) \quad (1.449)$$

$$= \left(\frac{1}{n^2}\right)^{1/3}\left(\sin G_1 \frac{m_2 m_1}{r_{21}^2}((R) + (\rho)) + (-\sin)G_1 \frac{m_2 m_1}{r_{21}^2}((R) + (\rho))\right)$$

the for

$$R_r = R\left(+\cos U \cos \omega + \sin U \sin \omega \cos \imath\right) \quad (1.450)$$

Proposition XXXXVIII (continued): Corollary VII - the rotation of the line of apsides

In Corollaries *VII* and *VIII* we returned to the problem of the rotation of the line of apsides we had considered earlier in Proposition *XXVI*.

Cor. *VII*. Given the axis of the ellipse described by the body P, or the line of apsides, does as to its angular motion go forwards and backwards by turns, but more forwards than backwards, and its direct motion in on the whole carried forwards.

We begin with the equation

$$\varpi = \omega + \Omega = \angle(e, \omega) + \angle(\omega, i) \quad (1.451)$$

when the inclination is set to zero we will then have

$$\varpi = \angle(e, i) \tag{1.452}$$

Since i is a fixed vector, $d\varpi / dt$ describes the rotation of the Lenz vector with respect to the fixed direction and therefore the rotation of the line of apsides. Then from the section, variation of the elements, for $\iota = 0$,

$$e\frac{d\varpi}{dt} = \left(\pm N_\alpha r^2 \left(F_S \oplus F_E\right) \cos \theta r\right) = \left(\pm N_\alpha r^2 \left(F_S \oplus F_E\right) \cos \theta \left(x_y e_1 + x_y e_2 + x_z e_3\right)\right) \tag{1.453}$$

which for substituting for F_r from Ch 11. we will have

$$
\begin{aligned}
e\frac{d\varpi}{dt} &= \frac{r}{a/R^2}\sqrt{\left(1 - e^2/R^2\right)}\left(\pm N_\alpha r^2 \left(F_S \oplus F_E\right)\cos\theta r\right) \\
&= \frac{r}{a/R^2}\sqrt{\left(1 - e^2/R^2\right)}\left(\pm N_\alpha r^2 \left(F_S \oplus F_E\right)\cos\theta \left(x_y e_1 + x_y e_2 + x_z e_3\right)\right)
\end{aligned} \tag{1.454}
$$

Then at the quadratures $(\psi = \pi/2 : 3\pi/2)$ and syzygies $(\psi = 0 : \pi)$ we will have then

$$
\begin{aligned}
e\left(\frac{d\varpi}{dt}\right)_0 &= \frac{r}{a/R^2}\sqrt{\left(1 - e^2/R^2\right)}\cos\varphi\left(\pm N_\alpha r^2 \left(F_S \oplus F_E\right)\cos\theta r\right) \\
&= \frac{r}{a/R^2}\sqrt{\left(1 - e^2/R^2\right)}\cos\varphi\left(\pm N_\alpha r^2 \left(F_S \oplus F_E\right)\cos\theta \left(x_y e_1 + x_y e_2 + x_z e_3\right)\right) \\
e\left(\frac{d\varpi}{dt}\right)_1 &= \frac{r}{a/R^2}\sqrt{\left(1 - e^2/R^2\right)}\cos\varphi\left(\pm N_\alpha r^2 \left(F_S \oplus F_E\right)\cos\theta r\right) \\
&= 2\frac{r}{a/R^2}\sqrt{\left(1 - e^2/R^2\right)}\cos\varphi\left(\pm N_\alpha r^2 \left(F_S \oplus F_E\right)\cos\theta \left(x_y e_1 + x_y e_2 + x_z e_3\right)\right)
\end{aligned} \tag{1.455}
$$

returning to equation (1.454) we find that on the contributions from the terms with the following actors $\cos 2\psi$ and $\sin 2\psi$ will vanish, we will have then for a secular term

$$
\begin{aligned}
e\langle d\varpi\rangle &= \sqrt{\left(1 - e^2\right)}\left\langle\left(\frac{r}{a}\cos\varphi\right)dt\right\rangle\left(\pm N_\alpha r^2 \left(F_S \oplus F_E\right)\cos\theta r\right) \\
&= \sqrt{\left(1 - e^2\right)}\left\langle\left(\frac{r}{a}\cos\varphi\right)dt\right\rangle\left(\pm N_\alpha r^2 \left(F_S \oplus F_E\right)\cos\theta \left(x_y e_1 + x_y e_2 + x_z e_3\right)\right)
\end{aligned} \tag{1.456}
$$

Using the equations in the context of Kepler's equation in Ch. 11, we will have then

$$\langle ed\varpi \rangle = \sqrt{(1-e^2)}\langle (\cos u - e)(1-e\cos u)\rangle \left(\pm N_\alpha r^2 \left(F_S \oplus F_E\right)r\right)$$
$$= \sqrt{(1-e^2)}\langle (\cos u - e)(1-e\cos u)\rangle \left(\pm N_\alpha r^2 \left(F_S \oplus F_E\right)\left(x_y e_1 + x_y e_2 + x_z e_3\right)\right)$$

(1.457)

where u denotes the eccentric anomaly. Since

$$\left(\cos u - e/R^2\right)\left(1-e/R^2\cos u\right) = \left(1+e/R^2\right)\cos u - e/R^2\left(1+\cos^2 u\right)$$
$$= -\frac{3}{2}e/R^2 + \left(1+e/R^2\right)\cos u - \frac{1}{2}e/R^2\cos 2u$$

(1.458)

we conclude then from equation (1.457)

$$\langle d\varpi \rangle = \sqrt{(1-e^2)}du\left(\sec^{-1}\right)\left(\pm N_\alpha r^2 \left(F_S \oplus F_E\right)r\right)$$
$$= \sqrt{(1-e^2)}du\left(\sec^{-1}\right)\left(\pm N_\alpha r^2 \left(F_S \oplus F_E\right)\left(x_y e_1 + x_y e_2 + x_z e_3\right)\right)$$

(1.459)

or

$$\langle d\varpi \rangle = \sqrt{(1-e^2/R^2)}u\left(\sec^{-1}\right)\left(\pm N_\alpha r^2 \left(F_S \oplus F_E\right)r\right)$$
$$= \sqrt{(1-e^2/R^2)}u\left(\sec^{-1}\right)\left(\pm N_\alpha r^2 \left(F_S \oplus F_E\right)\left(x_y e_1 + x_y e_2 + x_z e_3\right)\right)$$

(1.460)

the line of apsides will rotate forward, in the direction of rotation, with a mean amplitude

$$\sqrt{(1-e^2/R^2)}\left(\pm N_\alpha r^2 \left(F_S \oplus F_E\right)\right)$$
$$= \sqrt{(1-e^2/R^2)}\left(\pm N_\alpha \left(x_y e_1 + x_y e_2 + x_z e_3\right)^2 \left(F_S \oplus F_E\right)\right)$$

(1.461)

The total force, the $G.R.F.F.$, can be given as

$$\left(\pm N_\alpha \left(F_S \oplus F_E\right)r\right) = \left(\pm N_\alpha \left(F_S \oplus F_E\right)\left(x_x e_1 + x_y e_2 + x_z e_3\right)\right)$$

(1.462)

or by using standard definitions we will have

$$G.R.F.F. = \left(\pm N_\alpha r(t) \left(F_S \oplus F_E \right) \right) = \left(\pm N_\alpha \left(x_y(t)e_1 + x_y(t)e_2 + x_z(t)e_3 \right) \left(F_S \oplus F_E \right) \right) \qquad (1.463)$$

then by measuring r in units of the radius of the Moon we will have

$$G.R.F.F. = \left(\pm N_\alpha{}^2 r \left(F_S \oplus F_E \right) \right) = \left(\pm N_\alpha{}^2 \left(x_y e_1 + x_y e_2 + x_z e_3 \right) \left(F_S \oplus F_E \right) \right) \qquad (1.464)$$

From equation (1.464) we will have

$$
\begin{aligned}
(G.R.F.F.)_0 &= \left(\pm N_\alpha{}^2 r \left(F_S \oplus F_E \right) \right) = \left(\pm N_\alpha{}^2 \left(x_y e_1 + x_y e_2 + x_z e_3 \right) \left(F_S \oplus F_E \right) \right) \\
(G.R.F.F.)_1 &= \left(\pm N_\alpha N_\beta{}^2 r \left(F_S \oplus F_E \right) \right) = \left(\pm N_\alpha N_\beta{}^2 \left(x_y e_1 + x_y e_2 + x_z e_3 \right) \left(F_S \oplus F_E \right) \right)
\end{aligned}
\qquad (1.465)
$$

where

$$N_\alpha \neq N_\beta \qquad (1.466)$$

This force decreases less then and more than the square of the ratio of the distance at the quadratures and at syzygies.

In the following proposition (finding the motion of the apsides in orbits approaching very near to circles), the rotation of the line apsides under the $(G.R.F.F.)$

$$
\begin{aligned}
(G.R.F.F.) &= \left(\pm N_\alpha K K^2 r \left(F_S \oplus F_E \right) \right) = \left(\pm N_\alpha K K^2 \left(x_y e_1 + x_y e_2 + x_z e_3 \right) \left(F_S \oplus F_E \right) \right) \\
&= \left(\pm N_\alpha K c r \left(F_S \oplus F_E \right) \right) = \left(\pm N_\alpha K c \left(x_y e_1 + x_y e_2 + x_z e_3 \right) \left(F_S \oplus F_E \right) \right)
\end{aligned}
\qquad (1.467)
$$

then the angle of revolution between the apsides is given by

$$180 \ = \sqrt{\frac{1-c}{1-4c}} \quad 180 \left(1 + \frac{3}{2} c \right) \qquad (1.468)$$

therefore for $(G.R.F.F.)$ prevailing at quadratures and at syzygies, the angle of revolution is given by

$$180 \left(1 - \frac{3}{2} \eta^2 \right) : 180 \left(1 + \frac{3}{2} \eta^2 \right) \qquad (1.469)$$

then at the quadratures, the upper apse will go backwards while at syzygies the line of apsides will go forwards. Then the advance in the syzygies is significantly greater - as the regression in the quadratures - the line of apsides will go forward.

Between the places of the syzygies and quadratures the effect is one of forward motion. At the octants $(\psi = \pi/4)$ the centrifugal force is

$$K\left(\frac{1}{r^2} - \frac{1}{2}\eta^2 r\right) \tag{1.470}$$

Proposition XXXXVIII: Corollaries VIII-XVII

Corollary VIII. the formation of the eccentricity

We begin with the equation

$$\frac{de}{dt} = \left(\pm N_\alpha (F_S \oplus F_E) r\right) = \left(\pm N_\alpha (F_S \oplus F_E)(x(t)e_x + y(t)e_y + z(t)e_z)\right) \tag{1.471}$$

We then obtain

$$\frac{de}{dt} = \sqrt{(1 - e^2/R^2)} \frac{r}{a/R^2} \left(\pm N_\alpha (F_S \oplus F_E) r\right)$$
$$= \sqrt{(1 - e^2/R^2)} \frac{r}{a/R^2} \left(\pm N_\alpha (F_S \oplus F_E)(x(t)e_x + y(t)e_y + z(t)e_z)\right) \tag{1.472}$$

Given then the following equalities, by bringing out the dependence on v explicitly

$$\varphi = F_T(v): \psi = (1 + F_T)v \tag{1.473}$$

then

$$e - e_0 = \frac{1}{4}F_T\left\{\pm 2\cos(P - v) + \frac{3}{3 \pm 2F_T}\cos\left[(3 \pm 2F_T)P - v\right] + \frac{9}{1 \pm 2F_T}\cos\left[(1 \pm 2F_T)P + v\right]\right\} \tag{1.474}$$

then ignoring the $\eta's$ in the denominator ad reverting to the variables φ and ψ

$$e - e_0 = (F_T)\{-2\cos(\varphi) + \cos(2\psi + \varphi) + 9\cos(2\psi - \varphi)\} \qquad (1.475)$$

then to bring out the dependence on ψ explicitly, we substitute for φ

$$\varphi = \psi - \gamma \qquad (1.476)$$

where $\gamma = \angle(R, e)$ is considered to be a constant during the synodic month, then we find

$$e - e_0 = \frac{\mu}{r} + F\,R\{-2\cos(\varphi - \gamma) + \cos(3\psi - \gamma) + 9\cos(\psi + \gamma)\} \qquad (1.477)$$

we then obtain after some basic calculations

$$(e - e_0) = \frac{\mu}{r} + F\,R\left[(1 + \cos^2\psi)\cos\psi\cos\gamma - (2 + \sin^2\psi)\sin\psi\sin\gamma\right] \qquad (1.478)$$

Where $\mu = F_E : F = F_S$

Then

at the quadratures $(\psi = \pi/2): e - e_0 = \pm\left(\frac{\mu}{r} + F\,R\sin\gamma\right)$

and

at the syzygies $\quad (\psi = 0): e - e_0 = \left(\frac{\mu}{r} + F\,R\cos\gamma\right)$

The $(G.R.F.F.)$ is given as

$$G.R.F.F. = \pm(N_\alpha N_\beta)r(t)(K) = \pm(N_\alpha N_\beta)(x(t)e_1 + y(t)e_2 + z(t)e_3)(K)$$
$$= \pm(N_\alpha N_\beta)r(t)(F_E \oplus F_S) = \pm(N_\alpha N_\beta)(x(t)e_1 + y(t)e_2 + z(t)e_3)(F_E \oplus F_S) \qquad (1.479)$$

and the 'accession of the new forces'

at the quadratures $\qquad \pm F_T R r$

231

at the syzygies $\qquad \pm 2F_T Rr$

Corollary IX: the formation of the inclination

The inclination is the greatest of the all when the nodes are in the syzygies. In moving from the syzygies to the quadratures the inclination is diminished at each impulse of the body in question to the nodes: and becomes least of all when the nodes are in the quadratures, and the body in the syzygies; then it increases in the same manner as it did prior, and the following when the nodes arrive at the syzygies, returns to its former strength. This entire process is not any variation of the inclination but the natural state of the inclination due to the presence of both the mass of the Sun and the Earth.

The variational equation then may be given as

$$
\begin{aligned}
\frac{d\iota}{dt} &= \frac{r}{a/R^2 \sqrt{\left(1-e^2/R^2\right)}} \left(\pm N_\alpha \left(F_S \oplus F_E\right)r\right) \\
&= \frac{r}{a/R^2 \sqrt{\left(1-e^2/R^2\right)}} \left(\pm N_\alpha \left(F_S \oplus F_E\right)\left(x(t)e_x + y(t)e_y + z(t)e_z\right)\right)
\end{aligned}
$$

(1.480)

The inclination is periodic with a period of one-half a nodal year with an amplitude $(3\eta/8)=1\cdot6$ as measured in radians.

The dependence of $\frac{d\iota}{dt}$ on $\sin\iota$ is implied, and there will be no change in the inclination if ι were to vanish. The inclination is a total whole in regards to the entire summed force, positioned in relation to the orbital plane. The periodic dependence on the original inclination is on $-\sin U$ implies then that the effect stays the same when the nodes are in syzygies $(U=0)$ and is at its maximum at quadratures $(U=\pi/2)$

When the body is in the syzygies, the inclination is in its lowest state, but returns to its original state at the nest node. Depending on the trigonometric functions, the original values of the inclination will be at two separate phases in its lunar motion; then if the nodes are positioned after the quadratures, that is between C and A, D and B, the movement of the body P from either node, the normal node to the outer node, the inclination of the plane is proportionally diminished due to the both the presence of the Sun and the Earth. Then when it passes in the next 45 degrees to the next quadrature, the inclination is increased, and again is decrease 45 degrees later, due to the presence of both the Sun and the Earth.

Then $(\iota-\iota_0)$ diminishes in the interval $\pi/4 \leq U \leq 3\pi/4$ and then increases in the following period.

Corollary X: the formation of the direction of the ascending node

As given in Corollaries *VIII* and IX

we will have the initial equation

$$\sin\iota\frac{d\Omega}{dt}=\left(\pm N_\alpha r\left(F_S\oplus F_E\right)r\right)=\left(\pm N_\alpha r\left(F_S\oplus F_E\right)\left(x(t)e_x+y(t)e_y+z(t)e_z\right)\right) \quad (1.481)$$

and retaining only terms of zero order in *e* we will have simply

$$\frac{d\Omega}{dt}=\left(\pm N_\alpha r\left(F_S\oplus F_E\right)r\right)=\left(\pm N_\alpha r\left(F_S\oplus F_E\right)\left(x(t)e_x+y(t)e_y+z(t)e_z\right)\right) \quad (1.482)$$

The nodes will continue to recede in their passage from this node to the next. The nodes then when situated in the quadratures, recede continually and at the syzygies they recede more slowly, then being retrograde or stationary, will be carried backwards or made to recede in each revolution.

We begin with the equation

$$F_T=\left(\pm N_\alpha r\left(F_S\oplus F_E\right)r\right)=\left(\pm N_\alpha r\left(F_S\oplus F_E\right)\left(x(t)e_x+y(t)e_y+z(t)e_z\right)\right) \quad (1.483)$$

It follows then that at the conjunction $(\psi = 0)$ we have an increment of the amount $N_\alpha (F_T) r^2 / R$ while at the opposition $(\psi = \pi)$ we have a decrement, both fully represented by the overall function F_r. This comprises the original lunar motion. The Sun's presence as a contribution to the lunar motion, are odd and at the conjunction and at the opposition will have opposite signs. These values are primarily dependent on the magnitude of momentum of the center of gravity and the inverse of the square of the distance of the Earth from the center of gravity.

Further elaborations: Corollaries XII-XV

Like in the Corollaries *I-X*, we consider the factor GM / R^3 in the expressions for F_T in the equations for $\omega_t / n / e_t / n, \iota_t, / \sigma$ and Ω_t / σ

Corollary XII. In the expression for F_T derived earlier in Ch.11, the factor GM / R^3, representing the mean motion of the Sun, relative to the Earth. This equality is valid if and only if T describes an orbit relative to an 'immovable Sun'. If F_T is to represent the true mean motion of T around S, under their mutual gravitational related behavior, we will have by Proposition LIX

$$F_T \ (S \text{ fixed}) = F_T \text{ (under the mutual connection between T and S)} \times \frac{M_T}{M_T + M} \ (1.484)$$

Corollary *XII*. In Corollaries *VII* and *X* we have shown the terms in the mean motion of the line of apsides and the mean regression of the line of ascending nodes are, aside from the sign, the same.

We will then have, in terms of the line of apsides that rotates forwards with a mean amplitude,

$$\langle \varpi \rangle_{av} = \pm F_T (t)$$
$$\langle \Omega \rangle_{av} = \ F_T (t) \qquad (1.485)$$

and as in Cor. *VII* and Cor. *X* we conclude:

234

Therefore, the mean motion of the line of apsides will be in a ratio to the mean motion of the nodes; and both of these motions will be directly as the periodical time of the body P, and inversely as the square of the periodical time of the body T. The increase and decrease of the eccentricity and inclination of the orbit PAB is part of the normal motion of the moon, including that of the apsides and nodes.

Cor. *XIII*. The mean force by which the body T is retained in its orbit about S, is in part to the force with which the body P is retained in its orbit about T in a ratio compounded of the ratio of the radius ST to the radius PT, and the squared ratio of the periodical time of the body P about T to the periodical time of the body T about S, in direct proportion to the radius of T from S.

The force in which 'T is retained in its orbit about S' is

$$F\ \frac{M\ m_m}{R^2} \tag{1.486}$$

and the mean force by which the body P is retained in its orbit is

$$GM_T / r^2 \tag{1.487}$$

The ratio is then

$$\left(GM\ /R^3\right)R : \left(GM_T / r^3\right)r = \frac{N^2}{n^2} \tag{1.488}$$

Proposition XXXXVIII

(continued): Corollaries XIV-XVI

Here we focus our attention to and in the context of the perturbations of a Kepler orbit to the theory of the tides.

Cor. *XIV*. By the same laws in which the body P revolves about the body T, we may have many fluid bodies to move around T at equal distances from it; and being numerous they may all be contiguous to each other, to form a fluid annulus or ring, of a round figure and concentric to the body T; and several parts of this ring, their motions by the same laws as the body P, will draw nearer to the body T and move swifter in the conjunction and opposition of themselves and the body S, than in the quadratures. And the nodes of this ring or its intersections with the plane of the orbit of the body S or T, will rest at the syzygies; but also out of the syzygies will be carried backwards or in a retrograde motion, being faster in the quadratures and slower in the others. There will also be an inclination of this ring, and the axis will oscillate in each revolution, and when this revolution is complete, will return to the former arrangement. Having these laws laid out, we conclude the forces by which the tides are affected are due to the greater or lesser presence of the $G.R.F.F.$ radiation on the fluids themselves.

Given a continuous ring of particles, each of which revolves, like P, about T by the same laws directly related to the $G.R.F.F.$ radiation. The given sequence, as determined by the superior numerals in the text, are correlated with the following corollaries

1. Cor. V . $TA = r(1-x) < TC = r(1+x)$ The pertubing function; equation (1.429)/(1.439)

2. Cor. III . $V_1(\psi = 0) > V_0(\psi = \pi/2)$ (m is less than 1: given then that we have the conjunction and quadrature, given as the square of the velocity)

3. Cor. XI . (Forming of the direction of the ascending node, equation (1.483); $\langle \Omega \rangle_{av} = \pm N_\beta F_T t$

Cor. XV. Suppose the spherical body T, of matter that is not a fluid, and extends itself as far as a ring, and the channel were cut all round its circumference containing water; and this sphere revolves uniformly about its own axis in the same periodical time.

This water being accelerate and retarded by turns, as in the last Corollary, will be quicker at the syzygies, and slower at the quadratures, than the surface of the globe, and will ebb and flow in its channel according to the difference of the radiation pressure against and away from the surface of the fluid channel.

In the second part of the corollary, from Corollaries V and VI of the Laws of Motion, we find then that the conclusions prior reached will not be affected if all three bodies, S, T and P were moving in the same rectilinear motion.

In the last part of the Corollary, the principle $G.R.F.F.$'s that act on the water in the channel are the perturbations in the $G.R.F.F.$'s, directed towards and away from the Sun. These are given as

$$\left(\pm N_\alpha r\left(F_S \oplus F_E\right)r\right)=\left(\pm N_\alpha r\left(F_S \oplus F_E\right)\left(x(t)e_x + y(t)e_y + z(t)e_z\right)\right) \quad (1.489)$$

and

$$\left(\pm N_\sigma r\left(F_S \oplus F_E\right)r\right)=\left(\pm N_\sigma r\left(F_S \oplus F_E\right)\left(x(t)e_x + y(t)e_y + z(t)e_z\right)\right) \quad (1.490)$$

This force will retain the values

$$N_\alpha\left(F_E \oplus F_S\right)r(t)= N_\alpha\left(F_E \oplus F_S\right)\left(x(t)e_1 + y(t)e_2 + z(t)e_3\right) \text{ at quadratures}$$

and

$$N_\beta\left(F_E \oplus F_S\right)r(t)= N_\beta\left(F_E \oplus F_S\right)\left(x(t)e_1 + y(t)e_2 + z(t)e_3\right) \text{ at syzygies} \quad (1.491)$$

the pressure of the $G.R.F.F.$ radiation will push the water downwards at the quadratures and depress it as far as the syzygies.

The ring of fluids of Corollary *XIV* is coplanar with the orbital plane of P about T; and the same is the channel of Corollary *XV* lying in the equator.

Corollary *XV*. Here we see that if we apply Corollary *X* on the regression of the nodes and Corollary *IX* on the inclination. Then for the nodes and inclination we will have

$$\begin{aligned}
\frac{d\iota}{dt} &=\left(-N_\alpha\left(F_E \oplus F_S\right)\right)\times m_1\left(R\right)r(t) \\
&=\left(-N_\alpha\left(F_E \oplus F_S\right)\right)\times m_1\left(R\right)\left(x(t)e_1 + y(t)e_2 + z(t)e_3\right)
\end{aligned} \quad (1.492)$$

and

$$\frac{d\Omega}{dt} = \left(\pm N_\alpha \left(F_E \oplus F_S\right)\right) \times m_1 \left(R\right) r\left(t\right)$$
$$= \left(\pm N_\alpha \left(F_E \oplus F_S\right)\right) \times m_1 \left(R\right)\left(x\left(t\right)e_1 + y\left(t\right)e_2 + z\left(t\right)e_3\right)$$

(1.493)

then using trigonometry and integrating we will have

$$\iota - \iota_0 = N_\eta F_T \sin \tau \cos 2U$$

(1.494)

then

$\iota - \iota_0 \{$ is a maximum at syzygies $(U = 0)$

(1.495)

$\iota - \iota_0 \{$ is a minimum at quadratures $(U = \pi/2)$

(1.496)

$\iota - \iota_0 \{$ vanishes at octants $(U = \pi/4 : 3\pi/4)$

(1.497)

then when the nodes are at quadratures $(U = \pi/2)$

both $\left|\dfrac{d\Omega}{dt}\right|$ and $\left|\dfrac{d\iota}{dt}\right|$ are at the maximum

then at octants $(U = \pi/4 : 3\pi/4)$

$$\iota - \iota_0 = 0$$

(1.498)

Cor. XVI. Because the redundant matter in the equatorial regions of a globe causes the nodes to move backwards, then by increasing the matter the retrograde motion is likewise increased, by diminution is decreased and by removal quite ceases.

Apparently, the pressure at the bottom of the channel balances the vertical component F_r of the G.R.F.F. which leads to the shift in tides.

Cor. XVII. From the motion of the nodes we have the constitution of the globe. If the globes retain unalterable the poles, and the motion of the nodes is retrograde, there will be a redundancy of the matter near the equator. If the motion is direct there will then be a deficiency. Given a uniform spherical globe at rest in a free space, then by a given impulse made obliquely on its surface to be driven from its place, it will receive a motion both circular and straight. Since it makes no difference of any axes that pass through its center, nor has a given propensity for one axis or one situation of the axis than to any other, by its own force it will not alter its axis or the inclination of this axis. If the globe be impelled by some new impulse in the seemingly same part of its surface as it has before, and this tendency is not affected by whether it comes sooner or later, these impulses or tendencies will create the same motion in the sense as if they had been produced at the same time; that is if this motion of the globe had been produced by some force compounded of both forces, the two opposing directions, parallel, of the coming and going radiation. These forces or impulses, lead to a motion about an axis of a given inclination. If the second force or impulse were made upon any other place of the equator of the initial motion; and also if the impulse were made upon any place of in the equator of the motion which would be generated by the following impulse alone; also when both forces are made in any place, these forces will generate the same circular motion as if they were applied together, and at once, in the place of the intersections of the equators of those motions, which would be generated separately. A homogenous globe will then not hold onto various motions distinctly, but will unite all forces impressed upon it, and resolve them into one; revolving with a uniform motion about an given singular axis, with an inclination that is seemingly invariable. *The inclination of the axis, or the velocity of the rotation is dependent on the mangitude of the* G.R.F.F.. If the globe be divided into two hemispheres, by any plane passing through its center, and the center by which the force is directed, will urge each hemisphere separately, but will incline the globe to a side in regard to its motion round its axis.

Proposition XXXXVIV-XXXXXI

Here we will select a coordinate frame when considering a three-body problem, specifically S, P and T. The effect is least when the common center of gravity is of the three bodies is at rest. That is when the innermost body T is affected by the same laws as the rest are; and is greatest when the common center of the three is, by the decrease of the motion of the body T, is moving in any way, the effect is more so.

Cor. I If the smaller bodies revolve about the larger body, the orbits defined will be nearer to that of an ellipse, and the areas themselves will be more uniform. The bodies affect each other, that is the largest body affects the middle sized body and both of these affect the smallest body, by forces that are both inversely the square of their distances. Then if the focus of each orbit be placed in the common center of gravity of all of the interior bodies (the focus of the first and innermost body be placed in the center of gravity of the largest and innermost body; the focus of the second orbit be in the common center of gravity of the three innermost and so on). Then if the innermost body is at rest, it will become the common focus of all the given orbits.

Proposition *LI*. Theorem *I*

Given an arbitrary system of bodies, labeled A,B,C,D, and so on, if any of these bodies, as A affects according to the G.R.F.F. laws, then with such forces that are inversely proportional as the squares of the distances from the main affecting body, and another body, as B, likewise affects also the rest of the bodies, with forces that are also inversely as the squares of the distances from the affecting body, then the affecting forces of the bodies A and B will be to each other as those bodies A and B to which those forces are directly related to.

Cor. I. Then if each of the bodies of the given system A,B,C,D and so on, does singly affect by the G.R.F.F. laws, all the rest of the other bodies with given forces that are inversely as the squares of the distances from the main affecting body, these forces of all the bodies not including the least of the bodies, will affect each other as the bodies themselves.

Cor. II. If each of the bodies of the system A,B,C,D and so on, does affect all the rest by the G.R.F.F.'s that are either inversely or directly in a ratio of any power of any of the distances from the main affecting body; or which are determined by the distances from each of the affecting bodies according to the G.R.F.F. laws; then the forces related to those bodies are as the bodies themselves.

If we are given a set of bodies, A, B, C, D and so on, each of which affects all the rest in various different ways as defined by the laws of the G.R.F.F. 's and according to the common law of the distances separating each body in question. If P and Q are two of the previously defined bodies, considering the distance between both we will have the G.R.F.F. laws between both of them in relation to each other

$$\frac{d^2 r_P}{dt^2} = +G_Q \left(r_Q - r_P \right) F \left(\left| r_Q - r_P \right| \right) \tag{1.499}$$

where F is a 'universal' function of the absolute distance $\left| r_Q - r_P \right|$, between P and Q and G_Q is a positive numerical factor dependent on a scalar property of the body Q and also defined more generally as a universal constant. Then in relation to the above listed equation for two of the bodies X and Y we will have

$$\frac{d^2 r_X}{dt^2} = +G_Y \left(r_Y - r_X \right) F \left(\left| r_Y - r_X \right| \right)$$

$$\frac{d^2 r_Y}{dt^2} = +G_X \left(r_X - r_Y \right) F \left(\left| r_X - r_Y \right| \right) \tag{1.500}$$

then from these equations it follows then

$$\left| \frac{d^2 r_X}{dt^2} \right| : \left| \frac{d^2 r_X}{dt^2} \right| = G_{m_Y} : G_{m_X} \tag{1.501}$$

and by the third law of motion

$$\left| \frac{d^2 r_X}{dt^2} \right| : \left| \frac{d^2 r_Y}{dt^2} \right| = m_Y : m_X \tag{1.502}$$

In Corollary I we stated then that the bodies whose forces decrease as the square of the distance, are given as

$$F\left(\left|r_X - r_Y\right|\right) \propto \left|r_X - r_Y\right|^{-2} \tag{1.503}$$

Cor. *III*. In a system of bodies whose related forces decrease as the square of the distances, if the smaller body revolves about a larger body in ellipses, where their common focus is in the center of the larger body, and by radii drawn to the larger body, will describe areas proportional to the times exactly. The forces to each body, the larger to the smaller always, aside from mass blocking perturbations, will be very nearly in the ratio of the bodies. And also conversely. This is also taken from Cor. of Prop *XLVIII*, as compared with the first Corollary of this Proposition.

14

The superb theorems

Introduction

This is a Newtonian approximation of the surface-gravity fields, in which the force that attracts particles near its surface, within the defined rotational field, will be forced downward towards its surface by a force whose magnitude is proportional to the magnitude of its rotational velocity.

Propositions LII-LIV

Proposition *LII*. Theorem *I*

If on the spherical surface there are equal G.R.F.F. 's decreasing as the squares of the distances from those points, then a corpuscle placed within that sphere will not be affected by those forces.

Proposition *LIII*. Theorem *II*

If to the many points of a sphere there are equal G.R.F.F. forces decreasing as the squares of the distance from those singular points, there be given both the density of the sphere and the ration of the diameter to the distance of the body from its center; the force which affects the body is proportional to the semidiameter of the sphere.

Given two homogenous spheres of equal density and radii R_1 and R_2, the two smaller bodies P_1 and P_2 of equal masses at distances C_1P_1 and C_2P_2 which are in ratio of their radii

$$C_1 P_1 : C_2 P_2 = R_1 : R_2 \tag{1.504}$$

then let p_1 and p_2 be two 'equal' and even smaller masses (i.e. of equal infinitesimal volumes) in the two spheres at distances P_1 and P_2 that are also in the same ratio of the radii R_1 and R_2, then we will have

$$p_1 P_1 : p_2 P_2 = R_1 : R_2 \tag{1.505}$$

the gravitation interactions, given as F_1 and F_2 exerted by P_1 and P_2 on p_1 and p_2 are in the ratio

$$F_1 : F_2 = \left(P_1 p_1\right)^{-2} : \left(P_2 p_2\right)^{-2} = R_1^{-2} : R_2^{-2} \tag{1.506}$$

this ratio means that for every pair of similarly situated particles in the two spheres. The entire net influence F_{T_1} and F_{T_2}, of the whole spheres, will be directly as the number of bodies, N_1 and N_2 in the two spheres. Then by some given assumption, all the corpuscles in the two spheres are of equal finite volumes, given as

$$N_1 : N_2 = R_1^3 : R_2^3 \tag{1.507}$$

the total forces are given then as

$$F_{T_1} : F_{T_2} = N_1 F_1 : N_2 F_2 = R_1 : R_2 \tag{1.508}$$

Cor. 1. The gravitational behavior between one corpuscle towards several particles consisting of one singular sphere will be to the connected behavior of the other, towards as many related particles consisting of the other sphere in a direct ratio compounded of the ratio of the particles themselves as a whole, and as the square of the distances inversely. The particles collected are as the whole sphere, that is as the cubes of the diameters and the distances are as the diameters; and the first ratio related with the last ratio taken twice inversely, becomes the ratio of the one diameter to the other diameter.

The $G.R.F.F.$'s acting on P_1 and P_2, are directly as the distances from the center of gravity. Then the periodic times of the orbits about the spheres at distances are proportional to R_1 and R_2 are equal; and vice versa.

Cor. *II*. If to the several points of any two solids whatever, of similar figure and density, there are equal $G.R.F.F.$'s, decreasing as the square of the distance from those specific points, these forces, which affect these corpuscles, will be to each other as the diameters of the solids.

Proposition LIV-LVI

Proposition *LIV*. Theorem *III*

If there are several points related to a given sphere, there will be equal $G.R.F.F.$'s decreasing as the the square of the distances from those points, then a corpuscle placed within the sphere is affected by a force proportional to its distance from the center.

Given a homogenous sphere $ACBD$, each point of which is the center of the $G.R.F.F.$, decreasing as the square of the distance from it, we seek to find the net force acting on an interior point. We then draw a sphere, $PEQF$, passing through P and concentric to $ACBD$. By Proposition *LII* in the spherical shell, bounded by $ACBD$ and $PEQF$, we will have no effect on the body P, there then only remains the effect due to the interior sphere $PEQF$. By Proposition *LIII* it is as the distance PS.

Proposition *LV*. Theorem *IV*

A corpuscle situated within the sphere is affected by a force, a $G.R.F.F.$, inversely proportional to the square of the distance from the center.

The affecting force linked to any given homogenous spherical body, denoted as S, at an external point at a distance r_1 takes the following form

245

$$F = C(S) r_1^{-2} \tag{1.509}$$

where $C(S)$ is a scalar property characterized by the whole-body S. This can be rewritten as

$$F = C(S) r_1^{-2} \rightarrow F_{\alpha\beta\gamma} = F_{(ijk)_\kappa} \oplus F_{(ijk)_\tau} \tag{1.510}$$

If we apply this equation to two homogenous spherical bodies, S_1 and S_2, of radii R_1 and R_2. The forces at points, distant r_1 and r_2 from their respective centers are in the following ratio

$$\frac{F(S_1 : r_1)}{F(S_2 : r_2)} \rightarrow \frac{F_{\alpha\beta\gamma}(S_1) r_2^2}{F_{(\alpha\beta\gamma)_2}(S_2) r_1^2} \tag{1.511}$$

now if we let

$$r_1 : R_1 = r_2 : R_2 = \sigma \tag{1.512}$$

we will have

$$\frac{F_1(S_1; \sigma R_1)}{F_2(S_2; \sigma R_2)} \rightarrow \frac{F_{\alpha\beta\gamma}(S_1) R_2^2}{F_{(\alpha\beta\gamma)_2}(S_2) R_1^2} \tag{1.513}$$

But by Proposition *LIII* (equation (1.506)) we will have

$$\frac{F_1(S_1; \sigma R_1)}{F_2(S_2; \sigma R_2)} \rightarrow \frac{F_{\alpha\beta\gamma}(S_1) R_2^2}{F_{(\alpha\beta\gamma)_2}(S_2) R_1^2} = \frac{R_1}{R_2} \tag{1.514}$$

then by combining equations (1.513) and (1.514)

$$\frac{F_{\alpha\beta\gamma}(S_1) R_2^2}{F_{(\alpha\beta\gamma)_2}(S_2) R_1^2} = \frac{R_1}{R_2} : \frac{F_{\alpha\beta\gamma}(S_1)}{R_1^3} = \frac{F_{(\alpha\beta\gamma)_2}(S_2)}{R_2^3} \tag{1.515}$$

Since S_1 and S_2 are any two homogenous spheres we may write more generally

$$F = F_{\alpha\beta\gamma} R^3 / r^2 \tag{1.516}$$

Where $\Gamma_{\alpha\beta\delta}$ is a constant, for all homogenous spheres of the same density. Since matter of uniform diversity can be generally characterized by its density, denoted as ρ, we can write

$$F = C\frac{\rho R^3}{r^2} \tag{1.517}$$

where $C = $ constant.

Cor. *III*. If a corpuscle placed without a homogenous sphere is affected by a force inversely proportional to the square of the distance from the center, the combined force of the sphere constituents, consists of the above labeled net force linked with the presence of the sphere.

Proposition *LVI*. Theorem *V*

If to several points of a given sphere there are an associated net force, decreasing as the square of the distance from the sphere's center, then another smaller sphere will be affected by a G.R.F.F.*, inversely proportional to the square of the distance of the center.*

The combining of constituent particles comprises the overall net force associated with every large spherical object. Passive particles consist of the sphere that undergoes the affect exerted by another larger sphere, and active particles comprise the sphere in which is generating the G.R.F.F..

Given a homogenous sphere *S* consisting of active particles, all of which comprise the center of the G.R.F.F., decreasing as the square of the distance from it. In Proposition *LV*, such a sphere will act on a passive particle p, at an external point with a force proportional to

$$F = C\frac{\rho R^3}{r^2} \rightarrow \wp\frac{M}{r^2} \tag{1.518}$$

where M denotes the 'weight' of the sphere. Here r is the distance of p from the center.

Now consider another sphere the same in every regard to the first sphere. Its gravitational affect at the point C will be proportional to the net effect of the combined presence of every particle, comprising of the sphere as a whole. Given again as

$$\wp\,\frac{M}{D^2} \tag{1.519}$$

where D is the distance between the two centers. This force will act on the particles of the second sphere only as a whole and that the entire sphere, not just the singular particles, can be passive. Assuming the masses of the two spheres are the same, such that the whole of each sphere, comprising of singular particles, are themselves both passive and active regarding both spheres.

Proposition LVII

If spheres, however different (as to their density and field equations), in the same ratio onwards from the center to the circumference; but everywhere similar, at every given distance from the center, on all sides round about; then the gravitational effect from one sphere on the other decreases as the square of the distance of the body being affected.

Consider first, two homogeneous spherical shells of differing densities, both between two concentric spheres, or radii R_1 and R_2 and of radii r_1 and r_2. The spherical shells can be expressed as th difference of the homogenous spheres or radii R_1 and R_2 and masses M_2 and M_1 in one of them and of radii r_1 and r_2 and masses m_2 and m_1 in the other. Then the gravitational connection between the two spherical shells can be given as

$$F_{\alpha\beta}\left(\,_1,\,_2\right)=\,_1\left(M_2-M_1\right)\left(m_2-m_1\right)D^{-2} \tag{1.520}$$

where

$$F_{\alpha\beta} = F_{(ij)_u} \oplus F_{(ij)_v} \qquad (1.521)$$

To obtain the whole or entire influence of the larger sphere on the smaller sphere more exactly, we divide the spheres by n concentric spheres, including in one case a sphere S_0 at the center of mass ΔM_0 and spherical shells of masses ΔM_1, ΔM_2,... ΔM_n and in the different case a central sphere of mass δm_0 and spherical shells of masses $\delta m_1, \delta m_2, ..., \delta m_n$. We obtain the interaction between the two whole spheres, as divided by initially finding the influence of a particular shell ΔM_i of one of the spheres on each of the shells $\delta m_k \, (k = 0, 1, ..., n)$ of the other sphere.

Summing over k, and the summing the resulting form over all $i (= 0, ..., n)$ we find

$$F_{\alpha\beta}\left(\sum_{i=0}^{N}\Delta M_i : \sum_{i=0}^{N}\delta m_k\right) = \frac{1}{D^2}\left[\sum_{i=0}^{N}\Delta M_i\left(\sum_{i=0}^{N}\delta m_k\right)\right] \qquad (1.522)$$

Now letting the total number of concentric spheres, n, increase 'in infinitum', we will then obtain the limit

$$F_{\alpha\beta}(M,m) = \frac{Mm}{D^2} \qquad (1.523)$$

where M and m are the masses of the two spheres.

Corollaries I and II. If there are a number of spheres, the interaction between any two of them is the same, independent of the presence of the others. More specifically, if there are n spheres of masses ΔM_1, ΔM_2,... ΔM_n at relative distances R_{ij} between M_i and M_j; the interaction may be given as

$$F_{\alpha\beta}(M_i, M_j) = \frac{M_i M_j}{R_{ij}^2} \qquad (1.524)$$

where

$$(i \neq j) \tag{1.525}$$

$(i \neq j)$ for all distinct pairs (i, j).

Corollaries III and IV. The distinction between writing

$$F_{\alpha\beta}\left(M_i, M_j\right) \propto \frac{V_i V_j}{R_{ij}^2} : F_{\alpha\beta}\left(M_i, M_j\right) = \frac{M_i M_j}{R_{ij}^2} \tag{1.526}$$

where V_i and V_j are the volumes of the two spheres.

Corollary V. Here we follow in the distinction made between both 'active' and 'passive' particles.

Cor. VI. If two spheres as mentioned above, revolve about others at rest, each about each, and the distance between the centers of the quiescent and revolving bodies are proportional to the diameters of the quiescent bodies, the periodic times will be equal.

Cor. VII. If the periodic times are equal, the distances then will be proportional to the diameters.

In these corollaries we will find that regarding first Proposition Chapter 11, we will replace the point masses M_2 and M_1, by finite spheres of the kind considered in this specific proposition.

In the first Proposition Chapter 11, by comparing the elliptical orbit (of semimajor axis a_s) described by M_1 about a quiescent mass M_2 and the revolving elliptical orbit (of semimajor axis a_R) described by the same mass M_1, about the common center of gravity of M_1 and M_2; we show then for the equality of the of the periodic times

$$a_s : a_R = \sqrt[3]{M_2} : \sqrt[3]{(M_1 + M_2)} = \left(\frac{4}{3}\pi\rho_2\right)^{1/3} R_2 : \sqrt[3]{(M_1 + M_2)} \tag{1.527}$$

where ρ_2 and R_2 denote the mean density and the radius of M_2; and conversely.

Cor. *VIII*. Relating the motions of bodies about the foci of conic sections, will take place when the affecting sphere, of any spherical form and condition similar to the one described above, is placed in the focus.

Proposition LVIII

Having finished the consideration of the gravitational interaction between spherical bodies when the points of the sphere are centers of attraction with a force proportional to the inverse square of the distances from the center of the affecting bodies, we now move onto consider other forms of $G.R.F.F.$ interactions. The simplest case is when the law of $G.R.F.F.$ interaction is directly as the distance.

Proposition *LVIII*. Theorem *VI*

If two several points of spheres there are $G.R.F.F.$'s proportional to the distances of the points from the affecting bodies, then the compounded force with which the two spheres affect each other is as the distance between the center of the spheres.

Case 1. Let $AEBF$ be a sphere; S is the center; P is a corpuscle being affected; $PASB$ the axis of the sphere passing through the center of the corpuscle; EF, ef two planes cutting the sphere, and perpendicular to the axis, and equidistant, one on one side, the other on the other, then taking from the very center of one of the spheres, G and g, the intersections of the planes and the axis; and H any point in the plane EF. EF and ef are equally thin circular disks of thickness dz, along the axis PB. The length of the disc will determine the diameter of the sphere and thus its overall size and mass, considering the case of ordinary stars or massive stellar objects. The length of EF corresponds directly to the magnitude of the divergence and convergence of the stellar radiation that acts directly on the corpuscle P, as also the sum of the contributions that lead to these defined linear vector spaces, two-dimensional vector fields, of circular rings or radius $\varpi = HG$ and width $d\varpi$. The larger the ring and width, the larger the corresponding mass and output and input radiation. If the force runs in the direction of PG we will have

$$\rho F_{\alpha\beta}(2\pi\varpi)(z)PG \tag{1.528}$$

where $\Gamma_{\alpha\beta}$ is the constant of proportionality in the law of $G.R.F.F.$ Then integrating the previous expression over ϖ from 0 to GE we find the $G.R.F.F.$ in ratio of the disc EF

$$\rho F_{\alpha\beta}\left[\left(\pi(GE)^2\right)(z)\right]PG \tag{1.529}$$

and as the same above in terms of the disc ef

$$\rho F_{\alpha\beta}\left[\left(\pi(ge)^2\right)(z)\right]PG \tag{1.530}$$

Then the forces of all the planes in the sphere, equidistant on each side from the center of this sphere, are the same as the sum of those planes multiplied by the distance, PS, as the entire sphere and the distance PS.

The affect of the sphere on P is given as

the effect of the sphere on $P = F_{\alpha\beta}\rho(V)PS$ (1.531)

Case 2. Here we have the interaction of two homogenous spheres, S_1 and S_2, with centers, C_1 and C_2, and summing over all the particles, affecting the sum of all the particles of another body, is as the distance.

S_2 will affect the summation of all the particles of S_1 with the following force

$$\rho_1 F_{\alpha\beta}(V_1)C_2C_1 \tag{1.532}$$

Case 3. If the effect of the constituents of sphere S_2, all of the particles of it, on S_1 will then be given as

$$\left[\rho_1 F_{\alpha\beta}\left(V_1\right) \times \rho_2 F_{\alpha\beta}\left(V_2\right)\right] \cdot C_2 C_1 \qquad (1.533)$$

where V_1 and V_2 are the respective volumes of both spheres.

Lemma I and Propositions LVIV-LVI

Here we are concerned with the force exerted on a distant particle by a homogenous sphere according to the law of $G.R.F.F.$'s. We will be employing the use of Integral Calculus. Here an integral over two variables is reduced to one over one of them, by inverting the order of the integrations and evaluating the integral over the other one.

Here S is the center of the given sphere or radius a: P is the point of which the $G.R.F.F.$ emanates. This is at a distance from the sphere being acted upon from that distant point, given as $R = PS$, from the center. EE' is a chord that cuts the axis at right angles at D; and $z = PD$ is the distance of P from D. The range of z is given as $(R-a) \leq z \leq (R+a)$.

The affect on the given sphere whose size and gravitational magnitude is given by the length of the circular discs EDE', at various distances $z = PD$ from P. The overall length, of the result of the summing over the circular rings qDq' or radii ϖ, will be in the range $0 \leq \varpi \leq DE$.

Consider the effect on the sphere from a given point P, where we have two parameters that determine the magnitude of the force being exerted. We have the mass of the sphere being acted upon as determined by the length of the disc, EDE', and the length of the radii PD.

We begin by using the length $y = Pq = \sqrt{\left(z^2 + \varpi^2\right)}$, and the force exerted along this line, as a specific line of force.

Integrating we will then have the effect along this line force as

$$F_1 = 2\pi z \int_0^{qD} d\varpi\varpi f\left(\sqrt{\left(z^2 + \varpi^2\right)}\right)\cos\theta \qquad (1.534)$$

since

$$\cos\theta = z / \sqrt{\left(z^2 + \varpi^2\right)} \qquad (1.535)$$

then

$$F_2 = 2\pi z \int_0^{qD} d\varpi\varpi f\left(\sqrt{a^2 - R^2 + 2Rz}\right)\cos\theta \qquad (1.536)$$

followed by

$$F_3 = 2\pi z \int_0^{qD} d\varpi\varpi f\left(R - a\right)\cos\theta \qquad (1.537)$$

on the opposing end in terms of the trigonometric functions we will start with

$$F_4 = 2\pi z \int_0^{q'D} d\varpi\varpi f\left(z^2 + \left(-\varpi\right)^2\right)\cos\theta \qquad (1.538)$$

and

$$F_5 = 2\pi z \int_0^{E'D} d\varpi\varpi f \sqrt{\left(\sqrt{a^2 - R^2 + 2Rz}\right)}\cos\theta \qquad (1.539)$$

and in terms areas swept and represented by double integrals

$$F_1 = 2\pi \int_{R-a}^{R+a} z \int_0^{ED} d\varpi \frac{\varpi}{\sqrt{\left(z^2 + \varpi^2\right)}} f\left(\sqrt{\left(z^2 + \varpi^2\right)}\right) \qquad (1.540)$$

taking the following formula

$$y^2 = PE^2 = PD^2 + ED^2 = z^2 + a^2 - \left(R - z\right)^2 = a^2 - R^2 + 2rz \qquad (1.541)$$

we will have then

$$F_2 = 2\pi \int_{R-a}^{R+a} z \int_0^{ED} d(DE) \frac{DE}{\sqrt{\left(\sqrt{a^2 - R^2 + 2Rz}\right)}} f\left(\sqrt{\left(\sqrt{a^2 - R^2 + 2Rz}\right)}\right)$$

(1.542)

then

$$F_3 = 2\pi \int_{R-a}^{R+a} z \int_0^{ED} dED \frac{ED}{\sqrt{(R-a)}} f\left(\sqrt{(R-a)}\right)$$

(1.543)

and for the mirror image of the above sections we will have the following formulas

$$F_4 = 2\pi \int_{R-a}^{R+a} z \int_0^{ED} d(\varpi') \frac{\varpi'}{\sqrt{\left(z^2 + (\varpi')^2\right)}} f\left(\sqrt{\left(z^2 + (\varpi')^2\right)}\right)$$

$$= 2\pi \int_{R-a}^{R+a} z \int_0^{ED} d(-\varpi) \frac{-\varpi}{\sqrt{\left(z^2 + (-\varpi)^2\right)}} f\left(\sqrt{\left(z^2 + (-\varpi)^2\right)}\right)$$

(1.544)

and

$$F_5 = 2\pi \int_{R-a}^{R+a} z \int_0^{ED} d(DE)' \frac{(DE)'}{\sqrt{\left(\sqrt{a^2 - R^2 + 2Rz}\right)}} f\left(\sqrt{\left(\sqrt{a^2 - R^2 + 2Rz}\right)}\right)$$

$$= 2\pi \int_{R-a}^{R+a} z \int_0^{ED} d((-)DE) \frac{-(DE)}{\sqrt{\left(\sqrt{a^2 - R^2 + 2Rz}\right)}} f\left(\sqrt{\left(\sqrt{a^2 - R^2 + 2Rz}\right)}\right)$$

(1.545)

Now in the opposing directions, inward rather than outward, along a two-dimensional phase space, we will have

$$F_6 = 2\pi \int_{R-a}^{R+a} z \int_0^{ED} d(-)\varpi \frac{(-)\varpi}{\sqrt{\left(z^2 - \varpi^2\right)}} f\left(\sqrt{\left(z^2 - \varpi^2\right)}\right)$$

(1.546)

then

$$F_7 = 2\pi \int_{R-a}^{R+a} z \int_0^{ED} d\big((-)DE\big) \frac{(-)DE}{\sqrt{\big(\sqrt{a^2 - R^2 + 2Rz}\big)}} f\left(\sqrt{\big(\sqrt{a^2 - R + 2Rz}\big)}\right) \qquad (1.547)$$

followed by

$$F_8 = 2\pi \int_{R-a}^{R+a} z \int_0^{ED} dED \frac{ED}{\sqrt{(R+a)}} f\left(\sqrt{(R+a)}\right) \qquad (1.548)$$

Proposition *LVII*. Theorem *VI*

In a sphere described about the center S with a radius SA, if there be given SI, SA and SP, continually proportional, the effect on a corpuscle within the sphere in any place is not to the effect on the same object without the sphere in the place P. P, without the sphere, is in a ratio compounded of the square root of the ratio of IS, PS, the distances from the center and the square root of the ratio of the $G.R.F.F.$, in relation to the center in those places P and I.

Let the distances $R_1 \geq a : R_2 \leq a$, of two corpuscles places outside and within the sphere are related in the following manner as

$$R_1 R_2 = a^2 \qquad (1.549)$$

Then for the law of $G.R.F.F.$ interaction proportional to the nth power of the distance, the forces F_1 and F_2 are in the ratio

$$F_1 : F_2 = \sqrt{R_1^{n+1}} : \sqrt{R_2^{n+1}} \qquad (1.550)$$

here we outline the following proof.

For

$$R_2 = a^2 / R_1 \qquad (1.551)$$

the limits are given as

$$R_1 + a = \left(\sqrt{R_1} + \sqrt{R_2}\right)\sqrt{R_1} : a + R_2\left(\sqrt{R_1} + \sqrt{R_2}\right)\sqrt{R_2}$$
$$R_1 - a = \left(\sqrt{R_1} - \sqrt{R_2}\right)\sqrt{R_1} : a - R_2\left(\sqrt{R_1} - \sqrt{R_2}\right)\sqrt{R_2}$$

(1.552)

then besides

$$R_1^2 + a^2 = R_1\left(R_1 + R_2\right) : R_2^2 + a^2 = R_2\left(R_1 + R_2\right)$$

(1.553)

we have the following integrals

$$F_1 = \frac{\pi}{R_1^2} \int\limits_{R_1-a}^{R_1+a} \Gamma_{\alpha\beta}\left[\left(R_1^2 + a^2\right) - y^2\right]dy$$

(1.554)

and

$$F_1 = \frac{\pi}{R_1^2} \int\limits_{\left(\sqrt{R_1}-\sqrt{R_2}\right)\sqrt{R_2}}^{\left(\sqrt{R_1}+\sqrt{R_2}\right)\sqrt{R_1}} \Gamma_{\alpha\beta}\left[R_1\left(R_1 + R_2\right) - y^2\right]dy$$

(1.555)

then

$$F_1 = \frac{\pi}{R_2^2} \int\limits_{\left(\sqrt{R_1}-\sqrt{R_2}\right)\sqrt{R_2}}^{\left(\sqrt{R_1}+\sqrt{R_2}\right)\sqrt{R_1}} \left(f(y)(\beta)\right)\left[R_1\left(R_1 + R_2\right) - y^2\right]dy$$

(1.556)

then if we have

$$y = \kappa\sqrt{R_1}$$
$$y = \kappa\sqrt{R_2}$$

(1.557)

we will have then

$$F_1 = \frac{\pi}{R_1^2} \int\limits_{\left(\sqrt{R_1}-\sqrt{R_2}\right)\sqrt{R_2}}^{\left(\sqrt{R_1}+\sqrt{R_2}\right)\sqrt{R_1}} \Gamma_{\alpha\beta} \kappa \left[R_1\left(R_1+R_2\right) - \kappa^2 \right] d\kappa$$

$$F_1 = \frac{\pi}{R_2^2} \int\limits_{\left(\sqrt{R_1}-\sqrt{R_2}\right)\sqrt{R_2}}^{\left(\sqrt{R_1}+\sqrt{R_2}\right)\sqrt{R_1}} \left(f\left(\kappa\right)\left(\beta\right) \right) \kappa \left[R_1\left(R_1+R_2\right) - \kappa^2 \right] d\kappa$$

(1.558)

15

Prolegomenon

Phenomena

Phenomenon I

The circumjovial planets, by some radii drawn to Jupiter's center, describes areas proportional to the times of description; and their periodic times, the stars at rest, are as the 3/2th power of their distances from its center.

Phenomenon II

The circumsaturn planets, by radii drawn to Saturn's center, describe areas proportional to the times of description; and their periodic times, the stars at rest, are as the 3/2th power of their distance from its center as determined by the G.R.F.F.

Phenomenon III

The stars fixed in space, the periodic times of the five main planets, and of the Earth about the Sun, are as the 3/2th power of their mean distances from the Sun, as determined by the G.R.F.F.

Phenomenon IV

The Moon, by a radius drawn to the Earth's center, describes an area proportional to the times of description.

Propositions

Proposition *LVIII*. Theorem *I*

For every body, that moves in a curved line described in a plane, then by a radius drawn to a center body either at rest or moving along a given line, describes around that center body areas that are proportional to the given times as directed and governed by the G.R.F.F.*'s. All of this is in direct relation to that center point.*

Proposition *LVIV*. Theorem *II*

Every body, as by a radius to the center of another central body, however moved, will describe exact areas about that given center directly proportional to the given times, as dictated by the G.R.F.F. *as which that other body is affected.*

Proposition *LX*. Theorem *III*

The G.R.F.F. *of given bodies, which by equal motions describe various ellipses, in direct ratio as drawn by radii from the center of the same ellipses; are to each other as the squares of the arcs described in also equal times divided by the radii of the ellipses.*

Proposition *LXI*. Theorem *IV*

We seek to find the motion of the apsides in orbits defined as ellipses.

Cor. *I*. If the G.R.F.F. is as any power of the altitude that same power can be found from the actual motion of the apsides; that is taking the entire angular motion, in which the body returns to the same apse, as which the angular motion is of a singular revolution, as any number as $mA^{(nn/mm)-3}$ of the altitude A. The index is the power $(nn/mm)-3$.

Cor. *II*. If the body is affected by the G.R.F.F., which is inversely as the square of the altitude, will revolve in an ellipse whose focus is the center of the previously mentioned forces. A new and foreign force could be added or subtracted from this force, then the motion of the apsides due to the foreign force may be known.

Proposition *LXII*. Theorem *V*

If there are two bodies S and P, affecting each other, the larger on the smaller, with a force inversely proportional to the square of their distance, revolves about a common center of gravity; then the principle of axis of the ellipse in which either of the two bodies, as P moves about the other body S, will be directly to the principle axis of the ellipse. This is where the smaller body P may describe in the same periodic time, about the other body S, which here is fixed.

Proposition *LXIV*. Theorem *VI*

Bodies whose related forces decrease as the square of the distances from their centers, will move among themselves in ellipses; and also by a radii drawn to the foci will describe areas in their orbital revolutions nearly proportional to the times.

Cor. *I*. The gravitational perturbation will be greatest when gravitationally affecting parts of the system directly affecting the larger body, are not to each other inversely as the squares of the distances from the larger body; more so if the inequality of this proportion is greater than the inequality of the proportion of the distances from the larger body. The G.R.F.F. will be in parallel directions and equally causing no perturbations in the motions of the other parts of the system directly. When it acts unequally, it then causes a perturbation somewhere, which is either less or greater, as the inequality is greater or less. The presence of larger forces acting on some bodies, according to the G.R.F.F. laws, will alter the affecting situations among themselves. This so-labeled perturbation added to the other perturbation, due to the inequality and the inclination of the lines, makes the overall perturbation larger.

Cor. *II*. If the parts of the system move in ellipses without any evident perturbations, they will be affected by the presence of other bodies. This force, relatively weak or impressed nearly equally and in parallel directions, will be impressed upon all smaller bodies, in relation to a larger body.

Propositio *LXV*. Theorem *VII*

If too many points of a given sphere there are the G.R.F.F.'s *, decreasing as the square of the distances from the point, another sphere, similar to the original, will be affected by a similar force inversely proportional to the square of the distance of the centers.*

Proposition *LXVI*. Theorem *VIII*

If spheres, however different in density, in the same ratio from the center of the circumference; but elsewhere similar, at every given distance from the center, on all sides round about; the G.R.F.F. *of every point decreases as the square of the distance of the body being affected by these forces; then the entire* G.R.F.F. *in which one larger sphere affects the other is inversely proportional to the square of the distance of the centers.*

Proposition *LXVII*. Theorem *VIV*

If to the points of a given sphere, there are equal G.R.F.F.'s, *decreasing as the square of the distances from those points, there then will be given both the density of the sphere and the ratio of the diameter of the sphere to the distance of the corpuscle from its center: then the force with which the corpuscle is affected is proportional to the semidiameter of the larger sphere.*

Proposition *LXVIII*. Theorem *X*

If to several points of a given sphere there is exerted a G.R.F.F.*, decreasing as the square of the distances from the points, then a corpuscle placed within the sphere is affected by a frame-dependent rotational force proportional to its distance from the center.*

Proposition *LXVIX*. Theorem *XI*

If the density of a fluid is proportional to the compression, and its parts be force downwards by the frame-dependent rotational force inversely proportional to the squares of the distances from the center: if the distances are taken in some harmonic progression, the densities of the fluid at those given distances will be so in a geometrical progression.

Proposition *LXX*. Theorem *XII*

The areas by which revolving bodies describe by radii drawn to some immovable center of force do lie as well in the same immovable planes, are also proportional to the times in which they are describc, their motion.

Proposition *LXXI*. Theorem *XIII*

If a body revolves in a given ellipse, we seek to find the law of the G.R.F.F. *force by which the force is tending to an away from the focus of the ellipse.*

Proposition *LXXIII*. Theorem *XIV*

If a body moves in the perimeter of a parabola; we seek to find the law of the G.R.F.F. *field both tending towards and away from the focus of that figure.*

Cor. *I*. From the last three Propositions it follows then that if anybody P goes from a place P with a given arbitrary velocity in the direction of a right line PR, and at the same time is affected by the action of a *G.R.F.F.* force that is inversely proportional to the square of the distances from the places from the center, the body will then move in one of the conic sections, with its focus in the center of force, and conversely. The focus, the point of contact, and the singular position of the tangent, given, a singular conic section will be described, which at that point have some curvature. This curvature is given from the *G.R.F.F.* force and the momentum of the body is given; the two orbits, touching one another, can be described by the same force and the same velocity.

Proposition *LXXIV*. Theorem *XV*

If three bodies, two of which as the larger bodies have forces that decrease as the square of the distances, will affect the two smallest bodies, aside from the largest, will be to themselves as the square of the distances; and the two smallest will revolve one about the other and the larger about the largest body. Then the interior of the two revolving bodies, will be by some radii drawn to the innermost and greatest body, describe round that body areas more directly proportional to the times, and a figure the similar to that of an ellipse, with its focus in the point of intersection of the radii, the largest body will not be attracted by the smaller bodies and only very slightly affected by the presence of the smaller bodies.

Proposition *LXXV*. Theorem *XVI*

The G.R.F.F. laws of force being given, then the exterior body S does by radii drawn to the point O, the common center of gravity of the interior bodies P and T, describe round that center areas more proportional to the times, and an orbit that is more near to the form of an ordinary ellipse with its focus in that center, then it will describe about the innermost and largest body T by some radii drawn to that given body.

16

The universal law of gravitation

Introduction

Here we focus on the newly developed theory of the three generalized gravitational fields (discluding the surface-gravitational field) in terms of the original universal law of gravitation.

Propositions LXXVI-LXXVIII

In Propositions LXXVI, LXXVII and LXXVIII, the inference of the G.R.F.F.*'s, due to the presence of planets affecting smaller bodies, or satellites, and of the Sun affecting primary planets is based on the given description of equal areas in equal times, by the satellites about the planets and the primary planets about the Sun (Proposition LXXVII and LXXVIII) and the natural inference of their inverse-square law from the proportionality of the periods to the* $3/2$ *th power of the given mean distances from their respective centers (Proposition LXXVIV, Corollary VI). The 'quiescence of the aphelion points' and the rotation of the line apsides will result from the 'aberration' due to the inverse-square law of force (Proposition XLV and its related corollaries).*

Proposition *LXXVI*. Theorem *I*

The forces by which the circumjovial planets are continually drawn of from their original motions, and retained in their natural orbits, tend to and away from Jupiter's center; are inversely proportional to the square of the distances of the given places of those planets from that center.

Proposition *LXXVII*. Theorem *II*

The forces by which the primary planets are continually drawn of from their original rectilinear motions, and retained in their correct orbits, tend to and away from the Sun; are inversely as the squares of the distances of the places of those planets from the Sun's center.

Proposition *LXXVIII*. Theorem *III*

By the force in which the Moon is retained in its orbit, tends to and away from the Earth due the presence of both the Earth and the Sun simultaneously; is inversely as the square of the distance of its place from both the Earth's center and the Sun's center as well.

The $G.R.F.F.$ acting on the Moon, is due to the presence of both the Earth and the Sun. This force is both inversely as the square of the distance of its place from both the Earth's center and the Sun's center as well, considering then that the slow rotation of the Moon's apogee, equal to 3 3'in a single complete revolution. This rotation may be due to the presence of the Sun and the Earth, since both are required to find the overall orbital motion of the Moon.

Cor. *I* From Example 2 of Proposition *XXVI*, a body describing a revolving orbit will return to the same apse m times while the fixed orbit completes n revolutions, the $G.R.F.F.$ will take the following form

$$G.R.F.F. \propto r^{\left(n^2/m^2+F\right)-3}$$

(here we have replaced A in Corollary I, Proposition XXVI, by r) We will have

$$P = 360 \tag{1.559}$$

then

$$\frac{n^2}{m^2} - 3 = -\frac{29523}{14641} = -2\frac{4}{243} \tag{1.560}$$

the law of force is then given as

$$r^{-2-(4/243)} = r^{-2.0165} \tag{1.561}$$

Cor. *II*. We need to take into account the influence of both the Earth and the Sun when considering the motion of the Moon's apogee. We combine the presence of both the Earth and the Sun with a force that is according to an inverse-square law for both larger gravitating objects. Then the net gravitational force may be given in the following form

$$G.R.F.F. \propto \left(r^{-2} - cr\right) \tag{1.562}$$

where *c* is some constant. This force will produce an angle of rotation

$$180\sqrt{\frac{1-c}{1-4c}} \quad 180\left(1+\frac{3}{2}c\right) \tag{1.563}$$

in excess of 180. With the estimate

$$c = \frac{100}{35745} \tag{1.564}$$

The Moon's apogee will move forwards by $1\,31'28''$, during a single revolution, and the apse of the Moon is actually twice as fast.

Then to form a motion of the apse that is in agreement with what is observed we have

$$c = \frac{200}{35745} \quad 0.005595 \tag{1.565}$$

that is, as 1 to $178\frac{29}{40}$. Then the $G.R.F.F.$'s will be

$$G.R.F.F. \propto r^{-2.065} \propto r^{-2} - 0.005595r \tag{1.566}$$

in term of the observed motion of the Moon's apogee.

Cor. *III*. If we were to augment the *G.R.F.F.*, by which the Moon is retained in its orbit, first that is in the proportion of $177\frac{29}{40}$ to $178\frac{29}{40}$, then also in proportion of the square of the semidiameter of the Earth to the mean distance of the centers of the Moon and Earth, the arbitrary force due to the presence of the Moon in relation to the surface of the Earth, this force, descending to the Earth's surface, will increase inversely as the square of the given height.

Proposition LXXIX and the Moon test

Proposition *LXXIX*. Theorem *IV*

The moon is drawn to both towards and away from the Earth, by this G.R.F.F., is continually drawn away from a give rectilinear motion and retained in its proper orbit.

(a.) The circumference of the Earth is

$$1.232496 \times 10^8 \text{ Paris feet} \qquad (1.567)$$

where

$$\text{one French inch (Pouce)} = 2.7070 \qquad (1.568)$$

or an English inch is 2.5400 cm.

(b) The mean lunar day is

$$27^d 7^h 43^m = 39343 \text{ mintues} \qquad (1.569)$$

(c) The angular arc, $\delta\theta$, that the Moon describes in its orbit, per minute, is

$$\delta\theta = \frac{2\pi}{39343} \text{ radians} \qquad (1.570)$$

Then the distance that the Moon descends towards the Earth (BD) in one minute is

$$BD = \left[(60 \times r) \times \delta\theta \right] \times \frac{1}{2}\delta\theta \qquad (1.571)$$

where r is the radius of the Earth. This can also be given, using prior listed values, as

$$BD = \frac{1}{2}\left(60 \times \frac{1.232496 \times 10^8}{2\pi} \right) \times \left(\frac{2\pi}{3.9343 \times 10^4} \right) \qquad (1.572)$$

Then on the surface of the Earth the gravity, deemed the surface gravity, is 3600 times stronger. A body under this influence will fall the same distance in a single second instead. This is

$$15\frac{1}{120} \text{French feet} \qquad (1.573)$$

or in centimeters

$$\left(15\frac{1}{120} \times 12 \right) \times 2.7070 = 487.5 \, \text{cm} \qquad (1.574)$$

The value or magnitude of the descent of a falling body under the influence of the Earth's surface gravity will be directly proportional to the magnitude of the rotational velocity of the rotating Earth. This can then be labeled as a frame-dependent phenomenon, or more specifically in terms of a rotating frame-of-reference. Assuming this force to be constant, a body will descend in a time t, a distance given by the following formula

$$s = \frac{1}{2}gt^2 \qquad (1.575)$$

this will then become

$$s \rightarrow \frac{1}{2}gt^2\left(\sqrt{1-\frac{\omega^2}{\alpha^2}}\right) \tag{1.576}$$

where α is an arbitrarily large constant.

In one second the body will descend by the following amount

$$g = 2s = 975.0\,\mathrm{cm}\,\mathrm{sec}^{-2} \tag{1.577}$$

We now focus on the relationship between inertial and gravitational masses as established earlier. The period of oscillation, T, of a pendulum is directly related to its length, l, by

$$T^2 = \pi^2 l / g(\beta) \tag{1.578}$$

a relation also known by Huygens. Then for a seconds pendulum, that is a pendulum whose period of oscillation is one second, the length is given by

$$l = g(\beta)/\pi^2 \tag{1.579}$$

while by equation (1.578)

$$g(\beta) = 2s(\beta) \tag{1.580}$$

then in terms of *l* in place of *g* we will have

$$s = \frac{1}{2}l\pi^2 \tag{1.581}$$

where we have for the length

$$l = 3\,\text{Paris feet and } 7/10\,\text{of an inch} \tag{1.582}$$

using the value for the length we will have

$$s = \left(15 + \frac{1.16}{12}\right)\text{French feet} \tag{1.583}$$

then we conclude that from Newton, the force in which the Moon is held in its orbit then becomes, at the surface of the Earth, equal to the force by which we observe heavy bodies there.

Scholium

From the prior given rules, *1* and *2*, we will have the following demonstration of that previous Proposition. There are given several moons revolving about Earth, similar to the system of Jupiter or Saturn, the periodic times of these moons (by the augment of induction) is by the same law which Kepler found to obtain among the planets; then the given *G.R.F.F.*'s will be inversely as the squares of the distances from the center of the Earth. The *G.R.F.F.* would no longer hold when touching a part of the surface of the Earth, i.e. the mountains. This moon when deviated from the force holding it in its orbit around the Earth, will descend to the surface of the Earth. Regardless of the mass of the falling objects, the falling velocity will be the same and in direct proportion to the magnitude of the rotational velocity of the rotating body. There is a contrast and difference between both forces.

The emergence of the law of gravitation

In Proposition *LXXX* states that from Propositions *LXXVI-LXXIX*, some parts of the universal law of gravitation are eminent. This law will be fully evident in Proposition *LXXIX*.

Proposition *LXXX*. Theorem *V*

The circumjovial planets move to and away simultaneously from Jupiter; the circumsaturnal to and away from Saturn; the circumsolar to and away from the Sun; and by these related forces are drawn from a original rectilinear motions and retained in curvilinear orbits.

For the revolutions of the circumjovial planets about Jupiter, the circumsaturnal about Saturn, and of Mercury and Venus, are of the same force of that of the Moon about the Earth. These forces are both very similar and different from the forces that hold the circumsolar planets in their orbits about the Sun. The forces that dictate the motion of these revolutions are directed in direct relation of the center of Jupiter and Saturn. These are slightly different from the forces that hold the planets in their revolutions about the Sun, in direct relation to the center of the Sun.

Cor. *I*. The general force tending to all the planets, that is to and away from the center of the Sun, is different from the forces of all satellites tending to and away from the center of the revolving planets.

Cor. *II*. The central force which tends to and away simultaneously from the center of any planet is inversely as the square of the distance from that center.

Cor. *III*. All planets orbits are slightly deviated from their central revolutions by the close presence of other revolving planets.

Proposition LXXXI: the confirmation of the equality of inertial and the gravitational masses by astronomical data

Proposition *LXXXI*. Theorem *VI*

All given bodies move towards and away from any given planet; and the weights of the bodies in the same manner, at equal distances from the center of the planet, are proportional to the quantities in which they contain.

In this proposition we find a strong argument for the universality of the law of gravitation. That is the universal quantity of the inertial and the gravitational masses. We begin we a reevaluation of Newton's original synopses.

Bodies, regardless of mass, will descend to the Earth's surface with velocities equal to each other, that is they descend to the Earth's surface from equal height in equal times.

The original Moon test, from the section, the proportionality of mass and weight and the experiments on the pendulums, was based on the length of the seconds pendulum, in which was established the equality of inertial and gravitational masses.

If our terrestrial bodies are taken to the orbit of the Moon, then with the Moon, deprived of its original motion as to fall to the surface of the Earth, then in equal times they would describe equal spaces with the Moon. They are to the Moon, in terms of quantity of matter, as their weights to its weight. The Jupiter's moons perform their revolutions in times which are the $3/2$th power of the proportion of their distances from Jupiter's center. Their accelerations toward the center would be within the surface gravity of Jupiter in direct proportion to the magnitude of the rotational velocity of the rotating Jupiter. That is equal times in equal distances. In contrast, the bodies that were to move inward to the center of the Sun, the circumsolar planets would do so in a value inversely proportional to the square of the distance from the center of the Sun. Also in equal spaces to equal times.

For those bodies affected by the presence of the Sun, in proportion to the quantity of matter. If, at equal distances from the Sun, the satellites in proportion to the quantity of matter, will be affected by the presence of the Sun, by a force greater than Jupiter; then the distance between the centers of the Sun and of the satellite's orbit, will always be greater than the distance between the centers of the Sun and of Jupiter. This is almost as the square root of that proportion.

We begin with the original equations from Chapter 11 (Application pertaining to the lunar motion). Then it is evident that the masses m_k $(k=1,2,3)$ which happen on the left-hand sides of these equations are actually inertial masses while those on the right-hand side are the gravitational masses. We now consider the possibility that gravitational and inertial masses are different in the astronomical context, we distinguish them by labeling them as $m_k^{(i)}$ and $m_k^{(g)}$. We will then have

$$m_k^{(i)} = m_k^{(g)} / \delta_k \tag{1.584}$$

The general representational equations for the field equations may then be given
as

$$\frac{d^2 r_1}{dt^2} = G \frac{m_1 \delta_2}{r_{12}^3}(r_2 - r_1) + G \frac{m_1 \delta_3}{r_{13}^3}(r_3 - r_1)$$

$$\frac{d^2 r_2}{dt^2} = G \frac{m_2 \delta_1}{r_{21}^3}(r_1 - r_2) + G \frac{m_2 \delta_3}{r_{23}^3}(r_3 - r_2)$$

$$\frac{d^2 r_3}{dt^2} = G \frac{m_3 \delta_1}{r_{31}^3}(r_1 - r_3) + G \frac{m_3 \delta_2}{r_{32}^3}(r_2 - r_3)$$

(1.585)

the masses on the right-hand side are the gravitational masses, with removing the
distinguishing superscripts (g). Then the equation representing the motion of the
moon begin with the following equation

$$m \frac{d^2 r_1}{dt} = \begin{pmatrix} G_1 \frac{M\, m_1}{r_{12}^2} \delta\big((R)-(r)\big)\sin\varphi + G_2 \frac{M\, m_1}{r_{12}^2} \delta\big((R)-(r)\big)(\cos+\sin)\varphi_i \\ + G_3 \frac{M\, m_1}{r_{12}^2} \delta\big((R)-(r)\big)(\cos+\sin)\varphi_j \\ + G_4 \frac{M\, m_1}{r_{12}^2} \delta\big((R)-(r)\big)(\cos+\sin)\varphi_k \\ + G_5 \frac{M\, m_1}{r_{12}^2} \delta\big((R)-(r)\big)(\cos+\sin)\varphi_l \end{pmatrix}$$

(1.586)

$$+ \begin{pmatrix} -G_6 \frac{m_2 m_1}{r_{12}^2} \delta(R)\sin\varphi + (-)G_7 \frac{m_2 m_1}{r_{12}^2} \delta(R)(\cos+\sin)\varphi_i \\ +(-)G_8 \frac{m_2 m_1}{r_{12}^2} \delta(R)(\cos+\sin)\varphi_j + (-)G_9 \frac{m_2 m_1}{r_{12}^2} \delta(R)(\cos+\sin)\varphi_k \\ +(-)G_{10} \frac{m_2 m_1}{r_{12}^2} \delta(R)(\cos+\sin)\varphi_l \end{pmatrix}$$

and

274

$$m\frac{d^2 r_2}{dt} = \begin{pmatrix} (-)G_1 \frac{M\ m_1}{r_{12}^2}\delta\big((R)+(r)\big)\sin\varphi + (-)G_2 \frac{M\ m_1}{r_{12}^2}\delta\big((R)+(r)\big)(\cos+\sin)\varphi_i \\[2mm] +(-)G_3 \frac{M\ m_1}{r_{12}^2}\delta\big((R)+(r)\big)(\cos+\sin)\varphi_j \\[2mm] +(-)G_4 \frac{M\ m_1}{r_{12}^2}\delta\big((R)+(r)\big)(\cos+\sin)\varphi_k \\[2mm] +(-)G_5 \frac{M\ m_1}{r_{12}^2}\delta\big((R)+(r)\big)(\cos+\sin)\varphi_l \end{pmatrix} \quad (1.587)$$

$$+ \begin{pmatrix} G_6 \frac{m_2 m_1}{r_{12}^2}\delta(R)\sin\varphi + G_7 \frac{m_2 m_1}{r_{12}^2}\delta(R)(\cos+\sin)\varphi_i \\[2mm] +G_8 \frac{m_2 m_1}{r_{12}^2}\delta(R)(\cos+\sin)\varphi_j + G_9 \frac{m_2 m_1}{r_{12}^2}\delta(R)(\cos+\sin)\varphi_k \\[2mm] +G_{10} \frac{m_2 m_1}{r_{12}^2}\delta(R)(\cos+\sin)\varphi_l \end{pmatrix}$$

then

$$m\frac{d^2 r_3}{dt^2} = \left(G_1 \frac{M\ m_1}{r_{12}^2}\delta(R)(\cos+\sin)\varphi_i + G_2 \frac{M\ m_1}{r_{12}^2}\delta(R)(\cos+\sin)\varphi_j \right)$$
$$+ \left((-)G_3 \frac{M\ m_1}{r_{12}^2}\delta(R)(\cos+\sin)\varphi_i + (-)G_4 \frac{M\ m_1}{r_{12}^2}\delta(R)(\cos+\sin)\varphi_j \right) \quad (1.588)$$

and

$$m\frac{d^2 r_3}{dt^2} = \left((-)G_1 \frac{M\ m_1}{r_{12}^2}\delta(R)(\cos+\sin)\varphi_j + (-)G_2 \frac{M\ m_1}{r_{12}^2}\delta(R)(\cos+\sin)\varphi_i \right)$$
$$+ \left(G_3 \frac{M\ m_1}{r_{12}^2}\delta(R)(\cos+\sin)\varphi_j + G_4 \frac{M\ m_1}{r_{12}^2}\delta(R)(\cos+\sin)\varphi_i \right) \quad (1.589)$$

We will then have for the complete equation along the given radius vector *(r)*

$$F_T = \pm N_\alpha \left(F_S \oplus F_E \right)\left(\delta_2 r + N_\beta \left(\delta_2 - \delta_1 \right) \right) R\psi \quad (1.590)$$

which if take from the full equation for F_T , from Chapter 12, then we will have in full for the Earth-Moon system

$$F_E = (-\sin)G_1 \frac{m_2 m_1}{r_{21}^3}\delta(r_2 - r_1) + (\sin)G_2 \frac{m_2 m_1}{r_1^3}\delta_2(r_2 - r_1)$$

$$+ (-(\sin + \cos))G_3 \frac{m_2 m_1}{r_{21}^3}\delta(r_2 - r_1) + (\sin + \cos)G_4 \frac{m_2 m_1}{r_1^3}\delta_2(r_2 - r_1)$$

$$= (-\sin)G_1 \frac{m_2 m_1}{r_{21}^3}\delta((r_2 - r_1) + (r_3 + r_2)) + (\sin)G_2 \frac{m_2 m_1}{r_1^3}\delta_2((r_2 - r_1) + (r_3 + r_2))$$

$$+ (-(\sin + \cos))G_3 \frac{m_2 m_1}{r_{21}^3}\delta((r_2 - r_1) + (r_3 + r_2)) + (\sin + \cos)G_4 \frac{m_2 m_1}{r_1^3}\delta_2((r_2 - r_1) + (r_3 + r_2))$$

$$\quad (1.591)$$

$$= (-\sin)G_1 \frac{m_2 m_1}{r_{21}^3}\delta((R) - \rho) + (\sin)G_2 \frac{m_2 m_1}{r_1^3}\delta_2((R) + \rho)$$

$$+ (-(\sin + \cos))G_3 \frac{m_2 m_1}{r_{21}^3}\delta((R) - \rho) + (\sin + \cos)G_4 \frac{m_2 m_1}{r_1^3}\delta_2((R) + \rho)$$

And for the Sun-Moon system we will have

$$F_S = \left\{ \begin{array}{l} \left(\sin G_1 \frac{M\ m_2}{r^3}\delta(r_3 - r_1) + (-\sin)G_2 \frac{M\ m_2}{r^3}\delta(r_3 - r_1) \right) \\[2ex] + \left(G_3 \frac{M\ m_2}{r^3}\delta(r_3 - r_1) + G_4 \frac{M\ m_2}{r^3}\delta(r_3 - r_1) \right)(\cos + \sin)_j \\[2ex] + \left(G_5 \frac{M\ m_2}{r^3}\delta(r_3 - r_1) + G_6 \frac{M\ m_2}{r^3}\delta(r_3 - r_1) \right)(-)(\cos + \sin)_k \end{array} \right\}$$

$$= \left\{ \begin{array}{l} \left(\sin G_1 \frac{M\ m_2}{r^3}\delta((R + \rho)) + (-\sin)G_2 \frac{M\ m_2}{r^3}\delta((R - \rho)) \right) \\[2ex] + \left(G_3 \frac{M\ m_2}{r^3}(R) + G_4 \frac{M\ m_2}{r^3}\delta(R) \right)(\cos + \sin)_j \\[2ex] + \left(G_5 \frac{M\ m_2}{r^3}\delta(R) + G_6 \frac{M\ m_2}{r^3}\delta(R) \right)(-)(\cos + \sin)_k \end{array} \right\}$$

$$= \left\{ \begin{array}{l} \left(\sin G_1 \frac{M\ m_2}{r^3}\delta((r_3 - r_1) + (r_2 - r_1)) + (-\sin)G_2 \frac{M\ m_2}{r^3}\delta((r_3 - r_1) + (r_2 - r_3)) \right) \\[2ex] + \left(G_3 \frac{M\ m_2}{r^3}\delta(r_3 - r_1) + G_4 \frac{M\ m_2}{r^3}\delta(r_3 - r_1) \right)(\cos + \sin)_j \\[2ex] + \left(G_5 \frac{M\ m_2}{r^3}\delta(r_3 - r_1) + G_6 \frac{M\ m_2}{r^3}\delta(r_3 - r_1) \right)(-)(\cos + \sin)_k \end{array} \right\} \quad (1.592)$$

Then given as

$$F_S = \left\{ \begin{array}{l} \left[\begin{array}{l} \sin G_1 \dfrac{M\ m_2}{r^3} \delta_1 \left(R\cos + (r) \right) + N_\beta \left(\delta_2 - \delta_1 \right) R\cos\psi \\[2mm] + (-\sin) G_2 \dfrac{M\ m_2}{r^3} \delta_2 \left(R\cos + (r) \right) + N_\beta \left(\delta_2 - \delta_1 \right) R\cos\psi \end{array} \right] \\[8mm] + \left[\begin{array}{l} G_3 \dfrac{M\ m_2}{r^3} \delta_3 \left(R\cos + (r) \right) + N_\beta \left(\delta_2 - \delta_1 \right) R\cos\psi \\[2mm] + G_4 \dfrac{M\ m_2}{r^3} \delta_4 \left(R\cos + (r) \right) + N_\beta \left(\delta_2 - \delta_1 \right) R\cos\psi \end{array} \right] \left(\cos + \sin \right)_j \\[8mm] + \left[\begin{array}{l} G_5 \dfrac{M\ m_2}{r^3} \delta_5 \left(R\cos + (r) \right) + N_\beta \left(\delta_2 - \delta_1 \right) R\cos\psi + \\[2mm] G_6 \dfrac{M\ m_2}{r^3} \delta_6 \left(R\cos + (r) \right) + N_\beta \left(\delta_2 - \delta_1 \right) R\cos\psi \end{array} \right] (-)\left(\cos + \sin \right)_k \end{array} \right\} \qquad (1.593)$$

We see that the terms in $\left(\delta_2 - \delta_1 \right)$ are one order lower in r/R then the standard tidal terms. If we were to place a stringent upper limit on the terms $\left(\delta_2 - \delta_1 \right)$, in terms of the well defined tidal terms, we will have

$$\frac{R}{r}\left(\delta_2 - \delta_1 \right) \quad 1 \qquad (1.594)$$

Then for the Earth-Moon and Jupiter-Callisto systems we will have

$$\text{Earth-Sun distance / Earth-Moon distance} \quad 389 \qquad (1.595)$$

and

$$\left(\delta_2 - \delta_1 \right) \quad 2.57 \times 10^{-3} \qquad (1.596)$$

following with

$$\text{Jupiter-Sun distance / Jupiter-Callisto distance} \quad 383 \qquad (1.597)$$

with

$$\left(\delta_2 - \delta_1 \right) \quad 2.61 \times 10^{-3} \qquad (1.598)$$

As given in Chapter 11 we have all of the variation of the elements of the Kepler orbit. Here though we restrict ourselves to the variation of the major axis a.

$$\frac{d\bar{a}}{dt} = \frac{2\bar{a}}{rn} F_i \tag{1.599}$$

where

$$F_S = \left\{ \begin{array}{l} \left(\sin G_1 \frac{M\ m_2}{r^3} \delta_1 \left(R\cos + (r) \right) + (-\sin) G_2 \frac{M\ m_2}{r^3} \delta_2 \left(R\cos + (r) \right) \right) \\ + \left(G_3 \frac{M\ m_2}{r^3} \delta_3 \left(R\cos + (r) \right) + G_4 \frac{M\ m_2}{r^3} \delta_4 \left(R\cos + (r) \right) \right) (\cos + \sin)_{j_\alpha} \\ + \left(G_5 \frac{M\ m_2}{r^3} \delta_5 \left(R\cos + (r) \right) + G_6 \frac{M\ m_2}{r^3} \delta_6 \left(R\cos + (r) \right) \right) (-)(\cos + \sin)_{j_\beta} \\ + \left(G_7 \frac{M\ m_2}{r^3} \delta_3 \left(R\cos + (r) \right) + G_8 \frac{M\ m_2}{r^3} \delta_4 \left(R\cos + (r) \right) \right) (\cos + \sin)_{k_\alpha} \\ + \left(G_9 \frac{M\ m_2}{r^3} \delta_5 \left(R\cos + (r) \right) + G_{10} \frac{M\ m_2}{r^3} \delta_6 \left(R\cos + (r) \right) \right) (-)(\cos + \sin)_{k_\beta} \end{array} \right\} \tag{1.600}$$

where

$$F_{S_r} = \left\{ \begin{array}{l} \left(\sin G_1 \frac{M\ m_2}{r^3} \delta_1 \left(R\cos + (r) \right) + (-\sin) G_2 \frac{M\ m_2}{r^3} \delta_2 \left(R\cos + (r) \right) \right) \\ + \left(G_3 \frac{M\ m_2}{r^3} \delta_3 \left(R\cos + (r) \right) + G_4 \frac{M\ m_2}{r^3} \delta_4 \left(R\cos + (r) \right) \right) (\cos + \sin)_{j_\alpha} \\ + \left(G_5 \frac{M\ m_2}{r^3} \delta_5 \left(R\cos + (r) \right) + G_6 \frac{M\ m_2}{r^3} \delta_6 \left(R\cos + (r) \right) \right) (-)(\cos + \sin)_{j_\beta} \\ + \left(G_7 \frac{M\ m_2}{r^3} \delta_3 \left(R\cos + (r) \right) + G_8 \frac{M\ m_2}{r^3} \delta_4 \left(R\cos + (r) \right) \right) (\cos + \sin)_{k_\alpha} \\ + \left(G_9 \frac{M\ m_2}{r^3} \delta_5 \left(R\cos + (r) \right) + G_{10} \frac{M\ m_2}{r^3} \delta_6 \left(R\cos + (r) \right) \right) (-)(\cos + \sin)_{k_\beta} \end{array} \right\} \tag{1.601}$$

By combining the above two equations we will then have

$$\frac{d\bar{a}}{dt} = \frac{\pm 2\bar{a}_0}{r} F_T R \left(\delta_2 - \delta_1 \right) \cos \left(v - U \right) : F_T = \sum_N \left(G \frac{m_m M}{R^3} R \right) \oplus \sum_N \left(G \frac{m_E m_m}{r^3} r \right) \tag{1.602}$$

or since $F_T = \sum_N \left(G\frac{m_m M}{R^3} R \right) \oplus \sum_N \left(G\frac{m_E m_m}{r^3} r \right) : F_T dt = d\omega$

and

$$v \pm U = (1 \pm P)v \qquad (1.603)$$

we have

$$d\bar{a} = \frac{\pm 2\bar{a}_0}{1} F_T \frac{R}{r} dv (\delta_2 - \delta_1) \cos\left[(1 \pm P)v \right] dv \qquad (1.604)$$

where $\bar{a}_0$ represents the mean value of $\bar{a}$. Then by integrating the above equation we will have

$$\bar{a} - \bar{a}_0 = N\bar{a}_0 (F_T) \frac{R}{r} (\delta_2 - \delta_1) \cos\psi \qquad (1.605)$$

and

$$a - \bar{a} = a_0 \cdot (F_E)$$

$$\oplus \begin{pmatrix} \left(\left(G_1 \frac{M\ m_2}{r^3} (N\cos\psi + (r)) + (-)G_2 \frac{M\ m_2}{r^3} (N\cos\psi - (r)) \right)\sin\varphi \right) \\ + \left(\left(G_3 \frac{M\ m_2}{r^3} (N\cos\psi) + G_4 \frac{M\ m_2}{r^3} (N\cos\psi) \right)(\sin\varphi + \cos\varphi)_{j_\alpha} \right) \\ + \left(G_5 \frac{M\ m_2}{r^3} (N\cos\psi) + G_6 \frac{M\ m_2}{r^3} (N\cos\psi) \right)(-)(\sin\varphi + \cos\varphi)_{j_\beta} \\ + \left(\left(G_7 \frac{M\ m_2}{r^3} (N\cos\psi) + G_8 \frac{M\ m_2}{r^3} (N\cos\psi) \right)(\sin\varphi + \cos\varphi)_{k_\alpha} \right) \\ + \left(G_9 \frac{M\ m_2}{r^3} (N\cos\psi) + G_{10} \frac{M\ m_2}{r^3} (N\cos\psi) \right)(-)(\sin\varphi + \cos\varphi)_{k_\beta} \end{pmatrix} \qquad (1.606)$$

From equation (1.604) if follows then that the term $(\delta_2 - \delta_1)$ is not to cover the tidal term. Where

$$G\frac{m_2 m_1}{r_{21}^3}(r_2 - r_1) = G_2 \frac{m_2 m_1}{r_{12}^3}((r_3 - r_1) + (r_2 - r_3)) = G_2 \frac{m_2 m_1}{(r)_{22}^3}((R) + (\rho)) \qquad (1.607)$$

The corresponding equation in the related form of equation (1.605) is given as

$$a - a_0 = N a_0 \cdot \left[G_1 \frac{M\, m_2}{r^3} \left(\left(\frac{r}{R}\right)^3 \cos N_\beta \psi + (r) \right) + (-) G_2 \frac{M\, m_2}{r^3} \left(R2 \cos N_\beta \psi - (r) \right) \right] \sin \varphi$$

$$+ \left[\left(G_3 \frac{M\, m_2}{r^3} \left(\left(\frac{r}{R}\right)^3 \cos N_\beta \psi \right) + G_4 \frac{M\, m_2}{r^3} \left(\left(\frac{r}{R}\right)^3 \cos N_\beta \psi \right) \right) (\sin \varphi + \cos \varphi)_{j_\alpha} \right]$$

$$+ \left[\left(G_5 \frac{M\, m_2}{r^3} \left(\left(\frac{r}{R}\right)^3 \cos N_\beta \psi \right) + G_6 \frac{M\, m_2}{r^3} \left(\left(\frac{r}{R}\right)^3 \cos N_\beta \psi \right) \right) (-)(\sin \varphi + \cos \varphi)_{j_\beta} \right] \quad (1.608)$$

$$+ \left[\left(G_7 \frac{M\, m_2}{r^3} \left(\left(\frac{r}{R}\right)^3 \cos N_\beta \psi \right) + G_8 \frac{M\, m_2}{r^3} \left(\left(\frac{r}{R}\right)^3 \cos N_\beta \psi \right) \right) (\sin \varphi + \cos \varphi)_{k_\alpha} \right]$$

$$+ \left[G_9 \frac{M\, m_2}{r^3} \left(\left(\frac{r}{R}\right)^3 \cos N_\beta \psi \right) + G_{10} \frac{M\, m_2}{r^3} \left(\left(\frac{r}{R}\right)^3 \cos N_\beta \psi \right) \right] (-)(\sin \varphi + \cos \varphi)_{k_\beta}$$

And

$$\bar{a} - \bar{a}_0 = N \bar{a}_0 \left(F_T \right) \left(\frac{r}{R} \right)^2 (\delta_2 - \delta_1) \cos \psi \quad (1.609)$$

here we have the occurrence of the combination $(r/R)^2 (\delta_2 - \delta_1)$ in equation (1.608).

The question was posed by Newton, that what is the change in the satellite-Sun distance at conjunction that will compensate for the contribution arising from equation (1.608) with a non-vanishing, yet relatively small, $(\delta_2 - \delta_1)$.

First we consider the change in $a - a_0$ resulting from a small change Δr in r at $\psi = 0$. By equation (1.607) we will have

$$(\Delta)\, a - \bar{a} = a_0 \cdot F_E$$

$$\oplus \left| \begin{array}{c} \left(\left(G_1 \dfrac{M \; m_2}{r^3}\left(\dfrac{r^2}{R^3}\Delta r + (r) \right) + (-)G_2 \dfrac{M \; m_2}{r^3}\left(\dfrac{r^2}{R^3}\Delta r - (r) \right) \right) \sin\varphi \\[12pt] + \left(\left(G_3 \dfrac{M \; m_2}{r^3}\dfrac{r^2}{R^3}\Delta r + G_4 \dfrac{M \; m_2}{r^3}\dfrac{r^2}{R^3}\Delta r \right)\left(\sin\varphi + \cos\varphi \right)_{j_\alpha} \right) \\[12pt] + \left(G_5 \dfrac{M \; m_2}{r^3}\dfrac{r^2}{R^3}\Delta r + G_6 \dfrac{M \; m_2}{r^3}\dfrac{r^2}{R^3}\Delta r \right)(-)\left(\sin\varphi + \cos\varphi \right)_{j_\beta} \\[12pt] + \left(\left(G_7 \dfrac{M \; m_2}{r^3}\dfrac{r^2}{R^3}\Delta r + G_8 \dfrac{M \; m_2}{r^3}\dfrac{r^2}{R^3}\Delta r \right)\left(\sin\varphi + \cos\varphi \right)_{k_\alpha} \right) \\[12pt] + \left(G_9 \dfrac{M \; m_2}{r^3}\dfrac{r^2}{R^3}\Delta r + G_{10} \dfrac{M \; m_2}{r^3}\dfrac{r^2}{R^3}\Delta r \right)(-)\left(\sin\varphi + \cos\varphi \right)_{k_\beta} \end{array} \right| \qquad (1.610)$$

while the change in equation (1.608) is

$$a - \bar{a} = 2a_0 \cdot F_E$$

$$\oplus \left| \begin{array}{c} \left(G_1 \dfrac{M \; m_2}{r^3}\left(R(\delta_2 - \delta_1) + (r) \right) + (-)G_2 \dfrac{M \; m_2}{r^3}\left(R(\delta_2 - \delta_1) \right) \right) \sin\varphi \\[12pt] + \left(\left(G_3 \dfrac{M \; m_2}{r^3} R(\delta_2 - \delta_1) + G_4 \dfrac{M \; m_2}{r^3} R(\delta_2 - \delta_1) \right)\left(\sin\varphi + \cos\varphi \right)_{j_\alpha} \right) \\[12pt] + \left(G_5 \dfrac{M \; m_2}{r^3} R(\delta_2 - \delta_1) + G_6 \dfrac{M \; m_2}{r^3} R(\delta_2 - \delta_1) \right)(-)\left(\sin\varphi + \cos\varphi \right)_{j_\beta} \\[12pt] + \left(\left(G_7 \dfrac{M \; m_2}{r^3} R(\delta_2 - \delta_1) + G_8 \dfrac{M \; m_2}{r^3} R(\delta_2 - \delta_1) \right)\left(\sin\varphi + \cos\varphi \right)_{k_\alpha} \right) \\[12pt] + \left(G_9 \dfrac{M \; m_2}{r^3} R(\delta_2 - \delta_1) + G_{10} \dfrac{M \; m_2}{r^3} R(\delta_2 - \delta_1) \right)(-)\left(\sin\varphi + \cos\varphi \right)_{k_\beta} \end{array} \right| \qquad (1.611)$$

The net change from the two main causes can be given as

$$a - \bar{a} = 2a_0 \cdot F_E$$

$$\oplus \begin{pmatrix} \left(\left(G_1 \dfrac{M\ m_2}{r^3}\left(R(\delta_2 - \delta_1) + (r)\right) + (-)G_2 \dfrac{M\ m_2}{r^3}\left(R(\delta_2 - \delta_1)\right) \right) \sin\varphi \\[2ex] + \left(\left(G_3 \dfrac{M\ m_2}{r^3} R(\delta_2 - \delta_1) + G_4 \dfrac{M\ m_2}{r^3} R(\delta_2 - \delta_1) \right)(\sin\varphi + \cos\varphi)_{j_\alpha} \right) \\[2ex] + \left(G_5 \dfrac{M\ m_2}{r^3} R(\delta_2 - \delta_1) + G_6 \dfrac{M\ m_2}{r^3} R(\delta_2 - \delta_1) \right)(-)(\sin\varphi + \cos\varphi)_{j_\beta} \\[2ex] + \left(\left(G_7 \dfrac{M\ m_2}{r^3} R(\delta_2 - \delta_1) + G_8 \dfrac{M\ m_2}{r^3} R(\delta_2 - \delta_1) \right)(\sin\varphi + \cos\varphi)_{k_\alpha} \right) \\[2ex] + \left(G_9 \dfrac{M\ m_2}{r^3} R(\delta_2 - \delta_1) + G_{10} \dfrac{M\ m_2}{r^3} R(\delta_2 - \delta_1) \right)(-)(\sin\varphi + \cos\varphi)_{k_\beta} \end{pmatrix} \quad (1.612)$$

And then we will have

$$\Delta(a - a_0) + \bar{a} - \bar{a}_0 = a_0 \cdot F_E$$

$$\oplus \begin{pmatrix} \left(\begin{array}{l} G_1 \dfrac{M\ m_2}{r^3}\left(R\left[N\dfrac{\Delta r}{R} + N_\sigma(\delta_2 - \delta_1) \right] + (r) \right) \\[2ex] + (-)G_1 \dfrac{M\ m_2}{r^3}\left(R\left[N\dfrac{\Delta r}{R} + N_\sigma(\delta_2 - \delta_1) \right] - (r) \right) \end{array} \right) \sin\varphi \\[3ex] + \left(\left(G_3 \dfrac{M\ m_2}{r^3} R + G_4 \dfrac{M\ m_2}{r^3} R \right)\left[N\dfrac{\Delta r}{R} + N_\sigma(\delta_2 - \delta_1) \right](\sin\varphi + \cos\varphi)_{j_\alpha} \right) \\[3ex] + \left(G_5 \dfrac{M\ m_2}{r^3} R + G_6 \dfrac{M\ m_2}{r^3} R \right)\left[N\dfrac{\Delta r}{R} + N_\sigma(\delta_2 - \delta_1) \right](-)(\sin\varphi + \cos\varphi)_{j_\beta} \\[3ex] + \left(\left(G_7 \dfrac{M\ m_2}{r^3} R + G_8 \dfrac{M\ m_2}{r^3} R \right)\left[N\dfrac{\Delta r}{R} + N_\sigma(\delta_2 - \delta_1) \right](\sin\varphi + \cos\varphi)_{k_\alpha} \right) \\[3ex] + \left(G_9 \dfrac{M\ m_2}{r^3} R + G_{10} \dfrac{M\ m_2}{r^3} R \right)\left[N\dfrac{\Delta r}{R} + N_\sigma(\delta_2 - \delta_1) \right](-)(\sin\varphi + \cos\varphi) \end{pmatrix}_{k_\beta}$$

$$(1.613)$$

here on the right-hand side we have ignored the given notational difference in a_0 and $\bar{a}_0$. The net change as it vanishes is then given as

$$\delta_2 - \delta_1 + \frac{9}{4}\frac{\Delta r}{R} = 0 \tag{1.614}$$

Though at the conjunction $\psi = 0$,

$$\text{(satellite-Sun distance)}_{\psi=0} = \text{planet-Sun distance}$$

$$(= R)\text{-planet-satellite distance}(= r) \tag{1.615}$$

therefore

$$\Delta \text{(satellite-Sun distance)} = -\Delta r \tag{1.616}$$

the new solution as related to Newton's solution is given as

$$\delta_2 - \delta_1 = 2.25\Delta\text{(satellite-Sun distance)}_{\psi=0}/\text{ planet-Sun distance} \tag{1.617}$$

this is different from the original solution which is 2 replaced by 2.25, having no significance in this connection. Then if

$$\delta_2 - \delta_1 = \frac{1}{1000} \text{ then } \Delta\text{satellite-Sun distance / planet-Sun distance} = \frac{1}{2250}\tag{1.618}$$

Then for the Jupiter-Callisto system

$$\text{Jupiter-Callisto distance} = 1883 \times 10^3 \text{km} \tag{1.619}$$

and

$$\frac{1}{2250}\text{Jupiter-Sun distance} = 349 \times 10^3\text{km} \quad \frac{1}{6}\text{Jupiter-Callisto distance} \tag{1.620}$$

We continue with the reformulation of Newton's original statement:

Then if, at equal distances from the Sun, the $G.R.F.F.$, dictating the movement of an satellite as determined by the presence of the Sun, are either greater or less then the same force in direct relation to Jupiter's movement due to the presence of the Sun by one $\dfrac{1}{1000}$ part of the entire *G.R.F.F.*, the distance of the center of the satellite's orbit from the Sun would be either greater or less then the distance of Jupiter from the Sun by one $\dfrac{1}{2000}\left[\dfrac{1}{2250}\right]$ part of the entire distance. That is by a fifth (sixth) part of the distance of the utmost satellite from the center of Jupiter; the eccentricity of the orbit. Though the orbits of the satellites are concentric to Jupiter, the affect of the presence of the Sun on Jupiter and all of its satellites are equal.

In a similar statement, the weights of Saturn and its satellites in relation to the presence of the Sun, at equal distances from it, are as their quantities of matter; the weights of the Moon and of the Earth in relation to the position of the Sun, are either none or proportional to the masses of matter in which they contain.

Further speaking, the weights of all of the parts of every planet in relation to the presence of any other planet are one to another as the matter in the several parts; if some parts are more affected by the presence of another planet and others less, then for the quantity of matter, the entire planet, according to the parts of it that with which it most abounds, would be more or less moved in proportion to the quantity of matter in the entire planet. If we were to hypothetically raise the terrestrial bodies to the orbit of the Moon, to be compared with that body, then if the weights of those bodies were to the weights of the outer parts of the Moon as quantities of matter in the one and the other respectfully, but to the weights of the internal parts in a larger or smaller proportion, then similar to that the weights of those bodies would be to the weight of the entire Moon in a larger or smaller proportion in relation to what we have shown.

Cor. *I* The weights of bodies do not necessarily depend on their form; for if those weights were to be altered with the forms, they would either be larger or smaller, according to the variety of forms, in equal matter.

Cor. *II*. Universally all bodies, satellites, that move or orbit about the Earth, according to the laws of the special *G.R.F.F.*'s, the weights of all, at equal distances from the Earth's center, are as the quantities of matter which they contain. This is the quality of all bodies.

Cor. *III* . All spaces are not equally full.

Cor. *IV* . If all of the given solid bodies are of equal density, and are without pores, then a given void, space or vacuum must be determined. Given as their densities are then their inertias are in the proportion to their bulk.

Cor. *V* . The power of gravity is different from the power of magnetism.

Proposition LXXXII: The universal law of gravitation

Proposition *LXXXII*. Theorem *VII*

There is the power of the G.R.F.F., *pertaining to all bodies, proportional to the size and density of the object.*

All planets are affected by the presence of each other, as inversely as the square of the distance of places from the center of the planets. This force is directly proportional to the matter that they contain.

If all parts of one planet is affected by the presence of another planet; this force is to every part of the force of the entire object or body. And by the third law of motion, every planet is equally affected by the presence of the other planet, as that planet is by the presence of the latter planet. Both planets, A and B, will move towards each other due to the presence of each other, by shielding the general G.R.F.F., as determined by the presence of a singular star, in this case the Sun.

Cor. *I* . The G.R.F.F. affecting any given planet, that is its orbit, is compounded from and affecting each part of the planet equally. Magnetic and electric force are similar. By affecting the whole, the force is affecting all of the parts likewise.

Cor. *II* . The G.R.F.F. affecting several equal parts of any body is inversely as the square of the distance of places from these parts.

Proposition LXXXIII and LXXXIV: the implications of the 'superb theorems'

Proposition *LXXXIII*. Theorem *VIII*

In two given spheres, affecting each other by their presence according to the laws of the universal G.R.F.F.'s, the matter in places on all of the sides roundabout and equidistant from the centers is the same, the given weight of either sphere affecting motion of the other will be inversely as the square of the distance between their centers.

 The given G.R.F.F., affecting a whole planet, was compounded of the same force affecting all of its parts, as is to the inverse proportion of the square of the distances from those parts. The total force constitutes smaller parts comprised of the entire force given along a two-dimensional phase space with three main components in terms of its trigonometric functions in that same given space.

Cor. *I*. The principles given above in Proposition *LXXXIII*. Theorem *VIII* , when concerning the motion of bodies about the focus of the given conic sections, will take the place of the larger affecting sphere as placed in the focus, and the those bodies move about that sphere.

Cor. *II*. Those principles demonstrated above, when relating to the motions of bodies about the foci of given conic sections, will occur when the affecting sphere of any form, is placed in the focus.

Corollaries *I – IV*

Since the orbits of planets and satellites are elliptical in nature, we will have

$$\frac{2\pi}{T} = \left(\frac{\mu}{a(\gamma)^3} \right)^{1/2} \tag{1.621}$$

where $\mu = F_{\alpha\beta\sigma}M$, a is the semimajor axis, and T is the period. Then if a is the mean radius of the orbit, we will have

$$F_{\alpha\beta\sigma}M = 4\pi^2 \frac{a(\gamma)^3}{T^2} \qquad (1.622)$$

The same proportionality applies to the weight. Then we assume that the central object, the Sun or planets, is spherically symmetric and of radius R, then a different form of the equation (1.621) will be

$$\frac{4}{3}\pi\left[F_{\alpha\beta\sigma}\right]\bar{\rho}R^3 = 4\pi^2 a(\gamma)^3 T^{-2} \qquad (1.623)$$

where $\bar{\rho}$ is the mean density of the object. The equation (1.622) can then be written as

$$\bar{\rho} = \frac{3\pi}{\left[F_{\alpha\beta\sigma}\right]}\left(\frac{a(\gamma)}{R}\right)^3 \frac{1}{T^2} \propto \left(\frac{a(\gamma)}{R}\right)^3 T^{-2} \qquad (1.624)$$

The value of the $G.R.F.F.$, at any distance L from the center is given as

$$g_L = \frac{\left[F_{\alpha\beta\sigma}\right]M}{L^2} = \frac{4\pi^2}{L^2}\frac{a(\gamma)^3}{T^2} \qquad (1.625)$$

The value of the field equations at the surface of the object is given as

$$g_R = \frac{\left[F(r)\right](\beta)M}{R^2} = \frac{4}{3}\pi\left[F(r)\right](\beta)\bar{\rho}R \qquad (1.626)$$

The newly derived equations here can be applied to the Sun-Venus, Jupiter-Callisto, Saturn-Titan, and the Earth-Moon systems.

Cor. *III*. The smaller the planets are, other things being equal, in terms of greater density; the force of gravity at the surface due to the rotation being nearer to equality. Likewise, other things being equal, in terms of greater density, as they are closer to the Sun. Jupiter is denser than Saturn, and the Earth more dense than Jupiter; if the planets were to be placed at different distances from the Sun, then according to the degrees of density, they will either experience greater or lesser radiation from the Sun.

Propositions LXXXV-LXXXIX

Proposition *LXXXV*. Theorem *VIV*

The motions of the planets will naturally subsist an exceedingly long period of time according to the lifetime of the Sun.

Given a globe of water frozen into ice, and thus moving relatively freely in our air, in the same time that it would describe the length of the semidiameter, would thus lose by the resistance of the air, $\dfrac{1}{4586}$ part is of its motion; and the same proportion holds almost in all globes and moved with any given velocity.

At the direct height of 200 miles above the Earth, the air is more rare than at the surface of the Earth in the direct ratio of 30 to 0.0000000000003998, or as 7500000000000 to 1. Thus the planet Jupiter, revolving a in medium of the medium the $1000000th$ part on its motion in 1000000 years.

Then the celestial regions being void entirely of air and exhalations, the planets and comets meeting with resistance in these spaces will continue with their natural motion under the influence of the *G.R.F.F.*, for an relatively large period of time.

Hypothesis I

The center of the system of the planets is continually moving.

The Sun is the moving center of the planetary system.

Proposition *LXXXVI*. Theorem *X*

The common center of gravity of the Earth, the Sun and all of the planets, is constantly moving. The center is moving, in this case it is always moving, at a very near straight line.

The system of the world, which consists of the entire space of planetary phenomenon, the frame in which the center of gravity of the system is constantly moving. We assume the same principle when considering larger systems: the motions in a star cluster in regard to its center of gravity; the motions in the galaxy with respect to its center and so on ad infinitum.

Proposition *LXXXVII*. Theorem *XI*

That the Sun is constantly moving, it though never recedes mainly from the common center of gravity of all the planets.

We here consider the case in which the Sun's center oscillates.

Considering both the Sun-Jupiter and Sun-Saturn systems, we will have

(Center of Sun-Jupiter) distance:

$$778.3 \times 10^6 \, km$$

M /mass of Jupiter   1046

(Center of Sun-center of gravity Sun and Jupiter) distance
$$7.44 \times 10^5 \, km$$

(Center of Sun-Saturn) distance:
$$1427 \times 10^5 \, km$$

M /mass of Saturn

3503

(Center of Sun-center of gravity of Sun and Saturn) distance:

$4.07 \times 10^5 \, km$

Radius of Sun

$6.96 \times 10^5 \, km$

Despite the fact that the Earth and all of the planets were placed on side of the Sun, the distance of the common center of gravity of all from the center of the Sun would hardly add up to a singular diameter of the Sun. Aside from that, the distances of those defined centers will always be less; therefore, that singular center of gravity is in a constant state of motion as a singular momentum. Despite that motion, the Sun will not recede far from that center of gravity in the planetary system.

Cor. *I.* The common center of gravity of the Sun, the Earth and all of the planets is the center of the planetary system. The Earth and all of the planets motion are dictated by the presence of the Sun.

According to the gravitational powers generated due to the presence of the Sun, in relation to the planets, and the gravitational powers due to the presence of both the Sun and planets, in relation to the moons, are in continual orbital motion due to the generated *G.R.F.F.*'s. The planet's and moon's movable centers are taken from the movable center of the heliocentric system. If the larger body were to be placed in the center, affecting the movements of all other smaller bodies by its presence, and that this larger body, in this case the Sun, is moved, as a fixed point can be chosen such that the center of the Sun recedes least and from which it would then recede even less if this body were denser and greater in mass, and thus less apt to be moved even more.

Proposition *LXXXVIII.* Theorem *XII*

The planets move in ellipses, which have their common focus in the center of the Sun; and, by certain radii drawn towards and away from that center, they will describe areas in their orbital motion that are directly proportional to the times of that description.

We have discussed the related motions of the planets from the previously mentioned Phenomena. The weights of the planets directed both towards and away from the Sun, are inversely as the square of the distances from the Sun's center, and because of the motion of the Sun the planets orbits are ellipses with the Sun at the focus and would describe areas proportional to the times of that description, by which revolving bodies describe by radii drawn to some immovable center and some immovable planes in which they move. Also for an ellipse, the law of gravity, *G.R.F.F.,* tending to that focus. Since the effects of planets presence, in relation to one another when become near enough to another, as being relatively small, they may be neglected.

Proposition *LXXXVIX.* Theorem *XIII*

The aphelions and nodes of the orbits of the planets are fixed, and due directly to the motion of the Sun as the center of gravity of the planetary system.

The aphelions are immovable, and so are the planes of the orbits. The nature and magnitude of the aphelions are due to and In direct proportion to the magnitude of the motion of the Sun as the center of gravity. If the planes are fixed, then and nodes will be as well. The inequalities due to the mutual interactions of the planets and comets in their orbital revolutions are small enough to ignore.

Cor. *II.* The fixed stars are immovable or moving in relation to the motion of the planets. This includes the aphelions and nodes of the planets.

16

The figure of the Earth and of the planets

Introduction:

The affect of the G.R.F.F. on any parts of a body will to the same force on the whole as the matter of the part to the matter of the whole. And according to this rule, universally allows that all bodies are endowed with this principle of mutual G.R.F.F.. Here in this section we will focus on the theory of the tides and the figure of the Earth. Both are in direct relation due to the rotation of the Earth and the presence of both the Sun and the Moon.

Proposition LXXXX and the related historical background

Proposition *LXXXX*. Theorem *I*

The axes of the planets are less than the diameters drawn perpendicular to the axes.

The radiation pressure due to the emanating radiation from the Sun, blocked by the presence of the Moon, gives Earth its oblate spherical figure. This is the same for other planets as well. This effects the tides on Earth as well. By the diurnal revolution, this elliptical motion, the parts receding from the axis endeavor to ascend above the equator; there fore if the matter is in a fluid state, then by its ascent towards the equator it will enlarge the diameter there, then by its descent towards the poles it will shorten the axis. Then by this assumption, the diameter of Jupiter is found shorter between pole and pole than from east to west. By a similar case if the Earth were not about the equator than at the poles, the seas would then subside about the poles, and rising towards the equator all of which is directly proportional to the relation of how much radiation from the Sun strikes the Earth and how much is blocked due to the presence of the Moon, that is how close it is to the Earth.

If there were an equal amount of radiation striking the Earth on all sides and parts, the shape of the Earth would be that of a sphere. The rotation of the Earth gives it a slightly oblate shape.

Proposition LXXXXI: the method of the canals

Here we will focus on the original concepts developed by Newton and the mathematical formulas developed by Maclaurin.

(a) New method of the canals

We begin with

$$\epsilon = \text{equatorial radius - polar radius / the mean radius } (R) \qquad (1.627)$$

be the ellipticity and we will have

$$m = \text{centrifugal or rotational acceleration at the equator}$$

$$/ \text{ rotational acceleration at the surface} = \frac{\Omega^2 R}{\Omega M / R^2} = \frac{\Omega^2 R^3}{\Omega M} \qquad (1.628)$$

Newton used his method of canals of water in the following manner, though we will replace his original hypothesis with a newly developed theory. Given a hole of unit cross-section bored from a given point A, on the equator to the center of the Earth and a similar hole bored from a pole, Q, to the center; these canals then were constructed so they were filled with water. Considering then that the water in the canal would be in equilibrium, we conclude then that the weights of the equatorial and polar columns of water would be equal. Though, along the equator the acceleration due to the inward radiation pressure is diluted by the downward rotational acceleration; and because both of these induced accelerations, in a homogenous body, vary from the center linearly with the distance (by Proposition *LIV*). The dilution factor remains constant and retains its original value at the boundary, namely m 1.

If a denotes the equatorial radius, the weight of the equatorial column is given by:

$$\text{weight of equatorial column} = \frac{1}{2}\rho a g_{eq}\left(1 \pm F(\beta)\right) \tag{1.629}$$

where (β) is the gamma factor due to the rotational-dependent motion of the rotating Earth and ρ

is the density. Then if b denotes the polar radius

$$\text{weight of polar column} = \frac{1}{2}\rho b(\beta)_{pole} \tag{1.630}$$

And since the two weights must be equal

$$a g_{eq}\left(1 \pm F(\beta)\right) = b g_{pole} \tag{1.631}$$

This constitutes our analysis of the newly developed formulas for the original theory of canals. This is valid for any given spheroid, regardless of eccentricity. If the eccentricity is small, we will have

$$b = a(1 \pm \epsilon) \tag{1.632}$$

we will then have from equation (1.630)

$$\frac{g_{pole}}{g_{eq}} = \frac{a}{b}\left(1 + F(\beta)\right) = \frac{\left(1 \pm F(\beta)\right)}{1 - \epsilon}\ \left(1 + \epsilon \pm F(\beta)\right) \tag{1.633}$$

It remains though to determine the ratio from above.

From Maclaurin's formula for determining the finite eccentricity of an oblate spheroid by ignoring the gravitational potential factor, we will have

$$\Omega_{pole} = 4\pi a(\beta)\frac{\sqrt{\left(1-(e)^2\right)}}{(e)^3}\left[(e)-\left(\sin^{-1}(e)\right)\sqrt{\left(1-(e)^2\right)}\right] \qquad (1.634)$$

and then

$$\Omega_{eq} = 2\pi a(\beta)\frac{\sqrt{\left(1-(e)^2\right)}}{(e)^3}\left[\sin^{-1}(e)-(e)\sqrt{\left(1-(e)^2\right)}\right] \qquad (1.635)$$

From equations (1.633) and (1.634) we will have, for small eccentricity

$$\Omega_{pole} \quad \frac{4}{3}\pi a(\beta)\left(1-\frac{1}{10}(e)^2\right) \quad \frac{4}{3}\pi a(\beta)\left(1-\frac{1}{5}\in\right) \qquad (1.636)$$

and

$$\Omega_{eq} \quad \frac{4}{3}\pi a(\beta)\left(1-\frac{1}{5}(e)^2\right) \quad \frac{4}{3}\pi a(\beta)\left(1-\frac{2}{5}\in\right) \qquad (1.637)$$

since

$$(e)^2 = 2\in \qquad (1.638)$$

for $\in$ small.

From equations (1.632), (1.635) and (1.636) it follows

$$\frac{\Omega_{pole}}{\Omega_{eq}} = 1+\frac{1}{5}\in= 1+\in-m \qquad (1.639)$$

(b) Corollary

From equation (1.632) we found the ratio of $\in$ to m is related to $\Omega_{pole}/\Omega_{eq}$ related to $O(\in)$. We are then going to use the following relations

$$\Omega_{eq} = \frac{4}{3}\pi a(\beta)\left(1-\frac{2}{5}\epsilon\right)$$

$$\Omega_{pole} = \frac{4}{3}\pi a(\beta)\left(1-\frac{1}{5}\epsilon\right)$$

(1.640)

Newton figured the effect of an oblate spheroid at an external point along the minor axis. Thought this does not correspond for obtaining the values of the force of rotation at the equator.

Starting with $\Omega_{pole} = \frac{4}{3}\pi a(\beta)\left(1-\frac{1}{5}\epsilon\right)$ the integral which resolves this difficulty will be given as

$$F = 2b - \sqrt{\left(1-(e)^2\right)} \int_{R(\gamma)-b}^{R(\gamma)+b} \frac{y\,dy}{\sqrt{-(e)^2 y^2 + 2R(\gamma)y - \left(R(\gamma)^2 - b^2\right)}}$$

(1.641)

At the pole, where $R = b$ the above equation reduces to

$$F = 2b(\gamma) - \sqrt{\left(1-(e)^2\right)} \int^{2b(\gamma)} \frac{y\,dy}{\sqrt{\left(-(e)^2 y^2 + 2b(\gamma)y\right)}}$$

(1.642)

The evaluation of this integral is given as

$$F = 2\frac{b(\gamma)}{(e)^3}\left[(e) - \left(\sin^{-1}(e)\sqrt{\left(1-(e)^2\right)}\right)\right]$$

$$= 2a(\beta)\frac{\sqrt{\left(1-(e)^2\right)}}{(e)^3}\left[(e) - \left(\sin^{-1}(e)\sqrt{\left(1-(e)^2\right)}\right)\right]$$

(1.643)

The downward force, at an external point along the major axis of an oblate spheroid due to its rotation - the cross-sections normal to the axis are not circular but elliptical. We can find a new solution to this problem.

First we consider the case of a sphere or radius a. By contracting the scale along the y-axis uniformly by a factor of $(1-\epsilon)$, we can obtain an oblate spheroid. Then by expanding the scale along the x-axis by a factor $(1+\epsilon)$, we will have an prolate spheroid. Here the volume both diminished and enhanced in both singular cases, are equal and similarly disposed in regard to the sphere we can conclude according to Newton

$$\Omega_{eqatB}^{(ob)} = \frac{1}{2}\left[\Omega_{poleatA}^{(prolate)} + \Omega_{atB}^{(sphere)}\right] \quad \sqrt{\left[\Omega_{poleatA}^{(prolate)} \cdot \Omega_{atB}^{(sphere)}\right]} \tag{1.644}$$

But what about the affect of a prolate spheroid at an external point along its major axis? Since the cross-sections normal to the axis are circular, has an apparent adaptation. We will then have, in place of equation (1.640)

$$F = 2a - \int_{-a(\beta)}^{+a(\beta)} \frac{a-x}{\sqrt{(a-x)^2 + z^2}} dx \tag{1.645}$$

where

$$\frac{x^2}{a^2} + \frac{z^2}{c^2} = 1 : c^2 = a^2\left(1-(e)^2\right) \tag{1.646}$$

then letting

$$y = a - x \tag{1.647}$$

we find then successively

$$(a-x)^2 + z^2 = y^2 + c^2\left(1-\frac{x^2}{a^2}\right) = y^2 + c^2 - \frac{c^2}{a^2}(a-y)^2$$

$$= y^2 + c^2 - \frac{c^2}{a^2}\left(a^2 - 2ay + y^2\right) = y^2\left(1-\frac{c^2}{a^2}\right) + 2\frac{c^2}{a}y \tag{1.648}$$

$$= (e)^2 y^2 + 2a\left(1-(e)^2\right)y$$

Therefore

$$F = 2a - \int_0^{2a} \frac{y\,dy}{\sqrt{\left[2a\left(1-(e)^2\right)y+(e)^2 y^2\right]}} = 2a\frac{1-(e)^2}{(e)}\left[\frac{1}{2(e)}\log\frac{1+(e)}{1-(e)} - 1\right] \quad (1.649)$$

Then for small $\in$

$$F \quad 2a\frac{1-(e)^2}{(e)}\left[\frac{1}{3}(e)^2 + \frac{1}{5}(e)^4 + \frac{1}{7}(e)^6\right]$$

$$\frac{2a}{3}\left(1-(e)^2\right)\left(1+\frac{3}{5}(e)^2+...\right) \quad \frac{2a}{3}\left(1-\frac{2}{5}(e)^2\right)$$

$$(1.650)$$

(b) Redetermination of Newton's determination $\Omega_{pole}^{(ob)}/\Omega_{eq}^{(ob)}$ and $\in/m$

The successive steps to determine $\Omega_{pole}^{(ob)}/\Omega_{eq}^{(ob)}$ are given as

1. $\Omega_{eq}^{(oblate)} = \frac{1}{2}\left[\Omega_{poleatA}^{(prolate)} + \Omega_{(1)}^{(sphere)}\right] = \frac{2}{3}\pi a\left(1-\frac{4}{5}\in+1\right) = \frac{4}{3}\pi a\left(1-\frac{2}{5}\in\right)$

2. $\dfrac{\Omega_{pole}^{(oblate)}}{\Omega_{(1-\in)}^{(oblate)}} = \dfrac{1-\frac{1}{5}\in}{1-\in} = 1+\frac{4}{5}\in$

3. $\dfrac{\Omega_{(1)}^{(sphere)}}{\Omega_{eq}^{(oblate)}} = \dfrac{1}{1-\frac{2}{5}\in} = 1+\frac{2}{5}\in$

4. $\dfrac{\Omega_{(1-\in)}^{(sphere)}}{\Omega_{(1)}^{(sphere)}} = 1-\in$

$2\times3\times4 : \dfrac{\Omega_{pole}^{(oblate)}}{\Omega_{eq}^{(oblate)}} = 1+\frac{1}{5}\in$

If $APBQ$, represents the figure of the Earth, not spherical anymore, and generated by the rotation of an elliptical structure about the lesser axis PQ; and given

ACQqca, representing a canal full of water, reaching from the one pole *Qq* to the center *Cc*, and rising to the equator *Aa*; the weight of the water in the leg of the canal *ACca* will be to the weight of the water in the other leg *QCca* as 289 to 288. This is because of the centrifugal force, caused by the rotation of the Earth, sustains and takes off one of the 289 parts of the weight, in one leg, and the weight of 288 in the other, as sustains the rest. Though then if the matter of the Earth were all uniform, and without any motion, and its axis *PQ* were to the diameter *AB* as 100 to, the downward force in a place *Q* towards the surface of the Earth due directly to its rotation, the labeled centrifugal force, would be to the force in the same place *Q* towards the surface of a sphere described about the center *C* with the radius *PC* or *QC* as 126 to 125. In the same argument, the force due to the Earth's rotation, in a given place *A* towards the surface of the spheroid generated by its rotation of the ellipse *APBQ* about the axis *AB* is to the rotationally generated force in the same place *A* towards the surface of the sphere described about the center *C* with the radius *AC*, as 125 to 126. This force in a place *A* towards the surface of the Earth, is a mean proportional between these forces towards the surface of the spheroid and this sphere. Because the sphere has a diameter *PQ* diminished in a proportion of 101 to 100, is transformed into the figure of the Earth; and by this figure, having a third diameter perpendicular to the two diameters *AB* and *PQ* diminished in the same proportion, is converted in to the spheroid. The previously mentioned force in *A*, in either case, is diminished nearly in the same proportion. Therefore this force in *A* towards the surface of the sphere described about the center *C* with the radius *AC*, is to the force in *A* towards the surface of the Earth as 126 is to 125$\frac{1}{2}$. This same force in the place *Q* directed towards the surface of the sphere about the center *C* with the radius *QC*, is to the same force at *A* towards the surface of the sphere described about the center *C* with the radius *AC*, in the proportion to the diameters (by Prop. *LXXII*, Book *I*), that is, as 100 to 101. Therefore if we were to compound those three proportions 126 to 125, 126 is to 125$\frac{1}{2}$, and 100 to 101, into one, the force due to the rotation of the Earth in the place *Q*, directed towards the surface of the

Earth, will be to the same force in the place A towards the surface of the Earth, as $126 \cdot 126 \cdot 100$ to $125 \cdot 125^{1/2} \cdot 101$: or as 501 to 500.

Since this rotationally-dependent force, due to the Earth's rotation on it's axis, in either leg of the canal $ACca$ or $QCcq$, is as the distance of the places from the surface of the Earth, directed generally towards the Earth's center, those legs are then conceived to be divided by transverse, parallel, and equidistant surfaces, into parts proportional to the wholes, the weights of any number of parts in the one singular leg $ACca$, will be to the weights of the same number of parts in the other leg as their magnitudes and the accelerative forces due to the rotation of the Earth as a whole sphere conjointly, that is, as 100 to 101, and 500 to 501, or as 505 to 501. And therefore, if the rotational-dependent force affecting every part in the leg $ACca$, arising from this diurnal or rotational force, was to the weight of the same part as 4 to 505, so that from the weight of every part, conceived to be divided into 505 parts, the rotational force will affect those parts equally. Then the weights would remain equal in each leg, and thus the fluid would remain in equilibrium. But this force to every part is to the weight of the same part as 1 to 289; that is, this force, is $\dfrac{1}{289}$ parts thereof. Therefore, by the rule of proportion, if this force makes the height of the water in the leg $QCcq$, by $\dfrac{1}{100}$ part of its whole height, this same force $\dfrac{1}{289}$ will make the excess of the height in the leg $ACca$ only $\dfrac{1}{289}$ part of the height of the water in the other leg $QCcq$; and therefore the diameter of the Earth at the equator is to its diameter from pole to pole as 230 to 229. And since the mean semi diameter of the Earth, according to Picard's mensuration, is 3923.16 miles (5000 feet to a mile), the Earth will be higher at the equator than at the poles by approximately $85,472$ feet or roughly $17^{1/10}$ miles. And its height at the equator will be about $19,658,600$ feet, and at the poles $19,573,000$ feet.

Additional formulas

$$\frac{1}{2}a(\gamma)\Omega_{eq}\left(1\pm F(\beta)\right)=\frac{1}{2}b(\sigma)\Omega_{pole} \qquad (1.651)$$

therefore

$$\frac{\Omega_{polar}}{\Omega_{equator}}=\frac{a(\gamma)}{b(\sigma)}\left(1\pm F(\beta)\right):\frac{501}{500}\left(1\pm F(\beta)\right):\frac{505}{500} \qquad (1.652)$$

(c) Application to the figure of Jupiter

We move forwards here from the earlier Proposition concerning the figure of Jupiter.

The following relation, previously defined by Newton, $\in/m=\dfrac{5}{4}$, can be directly applied to any round object of known angular velocity of rotation and mean density, denoted ρ. Since then

$$m=\frac{\Omega^2 a(\gamma)}{\Omega M/a(\gamma)^2}=\frac{\Omega^2 a(\gamma)}{\frac{4}{3}\Omega\pi\bar{\rho}a(\gamma)^3/a(\gamma)^2}=\frac{3\Omega^2}{4\pi\Omega\bar{\rho}} \qquad (1.653)$$

another form of the expression for $\in$ can be given as

$$\in=\frac{15\Omega^2}{16\pi\Omega\bar{\rho}} \qquad (1.654)$$

accordingly, then

$$\in_{Jupiter}=\in_{Earth}\left(\frac{\Omega_{Jupiter}}{\Omega_{Earth}}\right)^2\frac{\bar{\rho}_{Earth}}{\bar{\rho}_{Jupiter}} \qquad (1.655)$$

Using the following data

Rotation period   Mean density $\left(g-cm^{-3}\right)$

Jupiter	$9^h 50^m$	$1 \cdot 314$
Earth	$23^h 56^m$	$5 \cdot 518$

we obtain then

$$\in_{jupiter} = \frac{1}{230}\left(\frac{23^h 56^m}{9^h 50^m}\right)^2 \frac{5 \cdot 518}{1 \cdot 314} \quad \frac{1}{9 \cdot 25} \tag{1.656}$$

here we have substituted $\in$ for the Earth as devised earlier. The above equation can then be given as

$$\in_{jupiter} \quad \frac{1}{230} \times \left(\frac{24}{10}\right)^2 \quad \frac{29}{5} \times \frac{400}{94\frac{1}{2}} \quad \frac{1}{9\frac{1}{3}} \tag{1.657}$$

here $400/94\frac{1}{2}$ is the ratio of the mean densities of Jupiter and the Earth, as we had derived earlier from Newton in the Proposition of the 'implications of the superb theorems'. Then the above equation can be given as

$$a(\gamma) : b(\beta) = 10\frac{1}{3} : 9\frac{1}{3} \tag{1.658}$$

and we can compare it with the range of values, $12:11 - 14\frac{1}{2} : 13\frac{1}{2}$, take from observations.

The variation of surface gravity over an oblate spheroid

Proposition LXXXXII. Problem I

We seek to find and compare the weights of bodies in different regions on the surface of the Earth.

302

Like in the Proposition, 'the method of the canals', we have shown that a slowly rotating homogenous body is a spheroid of small eccentricity, a value that is dependent of the magnitude of the body's rotation. We consider here the variation of this force, due to its rotation, over the surface of this spheroid. In regards to Newton, we focus on the generalization of the newly developed theory of the canals. Given a canal CA, of cross-section, bored radially outward from the center C of the spheroid to its boundary at A; and this canal filled with water of a given density, denoted σ, we have shown, originally from Newton, that the effect on the water due to the rotation of the spheroid varies linearly as the distance ξ from its surface. That is

$$(-f(r))(\beta) = C\xi \tag{1.659}$$

Where C is some constant. Then

$$\text{the weight of the water in the canal} = C\sigma \int_0^r \xi d\xi = \frac{1}{2}C\sigma r^2 \tag{1.660}$$

where r is the length, AC, of the canal. Since the value of the surface gravity at A is

$$(-f(r))\beta = C(r) \tag{1.661}$$

it follows then that

$$\text{the weight of the water in the canal} = \frac{1}{2}(-)f(r)\beta = C(r) \tag{1.662}$$

For the mutual equilibrium of all the canals, similarly bored, the necessary condition is that the resolved components of the weights of the canals normal to the boundary are then equal. Then

$$-f(r)\beta\cos\phi = K \tag{1.663}$$

where K is some constant; also given as

$$-f(r)\beta \cdot (p) = K \qquad (1.664)$$

where p is the perpendicular distance of C from the tangent at A.

It follows then

$$(-f(r))\beta - (-f(r))\beta_{eq} = K\left(\frac{1}{p} - \frac{1}{a(\gamma)}\right) \qquad (1.665)$$

where a denotes the semimajor axis of the spheroid.

From the equation of the tangent at (x_1, y_1): namely

$$\frac{xx_1}{a(\gamma)^2} + \frac{yy_1}{b(\beta)^2} = 1 \qquad (1.666)$$

then by a more familiar formula

$$
\begin{aligned}
\frac{1}{p^2} &= \frac{x^2}{a(\gamma)^4} + \frac{y^2}{b(\beta)^4} = \frac{x^2}{a(\gamma)^4} + \frac{1}{b(\beta)^2}\left(1 - \frac{x^2}{a(\gamma)^2}\right) \\
&= \frac{1}{b(\beta)^2}\left[1 - \frac{x^2}{a(\gamma)^4}\left(a(\gamma)^2 - b(\beta)^2\right)\right] = \frac{1}{a(\gamma)^2 b(\beta)^2}\left(a(\gamma)^2 - (e)^2 x^2\right)
\end{aligned}
\qquad (1.667)
$$

or

$$\frac{1}{p} = \frac{1}{a(\gamma)b(\beta)}\left(a(\gamma)^2 - (e)^2 x^2\right)^{1/2} \qquad (1.668)$$

Then

$$\frac{1}{p} - \frac{1}{a(\gamma)} = \frac{1}{a(\gamma)} \left[\frac{a(\gamma)}{b(\beta)} \left(1 - \frac{(e)^2 x^2}{a(\gamma)^2} \right)^{1/2} - 1 \right]$$

$$= \frac{1}{a(\gamma)} \left[\frac{1}{(1-(e)^2)^{1/2}} \left(1 - \frac{(e)^2 x^2}{a(\gamma)^2} \right)^{1/2} - 1 \right] \frac{(e)^2}{2a(\gamma)} \left(1 - \frac{x^2}{a(\gamma)^2} \right) = \frac{\epsilon}{a(\gamma)} \sin^2 \theta \qquad (1.669)$$

setting $x = a(\gamma) \cos \theta$.

We can conclude then for ϵ 1, the variation of the impact of surface gravity, defined by the rotation of the spheroid is then given by

$$\left(-f(r)\beta \right) - \left(-f(r)_{eq} \beta \right) \quad K \frac{\epsilon}{a(\gamma)} \sin^2 \theta \qquad (1.670)$$

Which is in agreement with the fact that the increase of weight in passing from the equator to the pole is as the square of the sine of the latitude.

To this Newton had added that the arcs of degrees of latitude in the meridian increase as nearly in the same proportion, meaning the variation of the curvature. The curvature of the orbit was given in Proposition *VI*.

We can write the equation of an ellipse in the form

$$y = \frac{b(\beta)}{a(\gamma)} \left(a(\gamma)^2 - x^2 \right)^{1/2} \qquad (1.671)$$

we find that

$$\frac{dy}{dx} = -\frac{b(\beta)x}{a(\gamma)\left(a(\gamma)^2 - x^2 \right)^{1/2}} ; \frac{d^2 y}{dx^2} = -\frac{a(\gamma)b(\beta)}{\left(a(\gamma)^2 - x^2 \right)^{3/2}} \qquad (1.672)$$

The expression for the curvature follows from its original definition; then

$$\rho = \left| \frac{\left[1 + (dy/dx)^2 \right]^{3/2}}{d^2 y/dx^2} \right| = \frac{1}{a(\gamma)b(\beta)} \left(a(\gamma)^2 - (e)^2 x^2 \right)^{3/2} - \frac{b(\beta)^2}{a(\gamma)} \qquad (1.673)$$

Then by equation (1.667)

$$\rho - \rho_{eq} = a(\gamma)^2 b(\beta)^2 \left(\frac{1}{p^3} - \frac{1}{a(\gamma)^3} \right) = \frac{1}{a(\gamma)b(\beta)} \left(a(\gamma)^2 - (e)^2 x^2 \right)^{3/2} - \frac{b(\beta)^2}{a(\gamma)}$$

$$= \frac{a(\gamma)^2}{b(\beta)} \left[\left(1^2 - (e)^2 x^2 \right)^{3/2} - \left(1 - (e)^2 \right)^{3/2} \right] \frac{3}{2} \frac{a(\gamma)^2 (e)^2}{b(\beta)} \sin^2 \theta = 3 \frac{a(\gamma)^2}{b(\beta)} \in \sin^2 \theta$$

(1.674)

or in a similar approximation

$$\rho - \rho_{eq} \quad 3a(\gamma) \in \sin^2 \theta$$

(1.675)

17

The lunar theory

Introduction

The basic elements of our newly developed formulation from Newton's theory has already been set out in the first seventeen corollaries or Proposition XXXXVIII. In the fourteen Propositions LXXXXIII-LLVI, the new theory is elaborated with its derivations and applications to the singular motion of the Moon about the Earth, due the presence of both the Earth and the Sun.

Propositions LXXXXIII and LXXXXIV

Proposition *LXXXXIII*. Theorem *I*

The diurnal motions of the planets are uniform, and the liberation of the Moon is due to this diurnal motion.

We have already considered the uniform motion of the planets in Chapter 3.

Due to the fact that the lunar day, arising from its uniform revolution about its axis, is menstrual, that is, equal to the time of its periodic revolution in its orbit, then the face of the Moon will always be nearly turned to the upper focus of its orbit. This situation of the focus will deviate a little to one side and to the other from the Earth in the lower focus; and this libration in longitude; then the libration in latitude arises from the Moon's latitude and the inclination of its axis to the plane of the ecliptic.

From the solution of Kepler's equation given earlier, in terms of finding the place of a body moving in a given ellipse at an given time, it is clear given Newton's formulation from tracing the longitudinal libration of the Moon to the difference $u - \varphi$, between the mean and the true anomalies. Using the notation given earlier in Chapter 11, equations (1.375-1.376), u and φ are defined by the following equations

$$r^2 \left(\left(F_{uv_\alpha} \oplus F_{uv_\beta} \right) \frac{Mm}{r^2} \right) = \text{the constant of areas} \qquad (1.676)$$

and then

$$[h] = \left[a(\gamma_\beta) \mu \left(1 - \{e\}^2 \right) \right]^{1/2} \qquad (1.677)$$

followed by the following relations

$$n = \left(\frac{\mu}{a(\gamma_\beta)^3} \right)^{1/2} = \left(1 - \{e\} \cos u \right) \frac{du}{dt} \qquad (1.678)$$

in a similar scenario the utmost satellite of Saturn will show the same face to Saturn.

Proposition LXXXXIV. Theorem II

The equinoctial points go backwards and the axis of the Earth, by a nutation in every annual revolution, does twice vibrates towards the ecliptic, as often returns to its original position.

The regression of the nodes (in conjunction with the formation of the inclination of the orbital plane) is considered in, the formation of the eccentricity and inclination, given in their related Propositions *XXXXVIII*.

Newton realized that the nutation θ is related, most likely, to the average precession of the equinoxes, by the proportionality

$$\theta = \langle \psi \rangle_{AV} \tan \theta \sin 2 (vt - \psi) \quad -50'' \tan \theta \sin 2 (vt - \psi)$$

Propositions LXXXXV and LXXXXVI

Proposition *LXXXXV*. Theorem *III*

All the motions of the Moon, and all the inequalities of those motions, follow from the principles which we have already laid down.

The greater planets, while they are moved about the Sun, may in the same time carry other lesser planets or satellites, revolving about the Sun, and that those lesser planets must move in ellipses which have their foci in the centers of the greater planets. But their motions will be in many ways must incorporate the action due to the presence of the Sun, and will suffer such inequalities as are observed in the Moon.

These elements, originally considered as perturbations, are considered here more extensively, Proposition XXXXVI: Corollary I-VI,

$$F = \left(\sum_N G\right)\frac{M\ m_m}{ST^2} = \left(\sum_N G\right)\frac{M\ m_m}{R^2} : F_E = \left(\sum_N G\right)\frac{m_E m_m}{ST^2} = \left(\sum_N G\right)\frac{m_E m_m}{R^2}$$

1. On the velocity along with the orbit: Corollary *III*, Prop. *XXXXVIII*

$$V_1 > V_0; \quad V_1 : V_0 = \frac{1+x}{1-x}\left(1 + \frac{3}{2}\frac{F_T}{1\pm P}\right)$$

2. On the constant of areas: Corollary *II*, Prop. *XXXXVIII*

$$[h_l] = \langle h \rangle_{av}\left(1 \pm \frac{3}{2}F_T\right) \quad \text{and} \quad [h_l] = \langle h \rangle_{av}\left(1\ \frac{3}{2}F_T\right)$$

3. On the curvature of the orbit: Corollary *IV*, Prop. *XXXXVIII*

$$\rho > \rho_0; \qquad \rho_1 : \rho_0 = 1 + 3(F_T)$$

4. On the prolateness of the orbit: Corollary *IV*, *XXXXVIII*

$$1 - x : 1 + x = 69/70$$

5. On the the eccentricity: Corollary IX , Prop. *XXXXX: Corollary VIII-XVII*

$$(e-e_0)_1 = 2(F_T)\cos\gamma : \quad (e-e_0)_0 = \pm 3(F_T)\sin\gamma$$

6. On the rotation of the line of the apsides: Corollary VII and VIII , Prop. *XXXXIX: continued Corollary VII (1.455)*

$$e\left(\frac{d\varpi}{dt}\right)_0 = \frac{\mu}{r} \pm \frac{r}{a}(1-e^2)^{1/2}(F_T)\cos\varphi$$

$$e\left(\frac{d\varpi}{dt}\right)_1 = \frac{\mu}{r} \; 2\frac{r}{a}(1-e^2)^{1/2}(F\;)\cos\varphi$$

but

$$\langle\varpi\rangle_{av} = \sqrt{(1-e^2)}(F_T).nt$$

7. On the regression of the nodes: Corollary XI , Prop. *XXXXX (1.482)*

$$\left(\frac{d\Omega}{dt}\right)_1 = \pm\frac{\mu}{r} + (F_T R)\sin^2 v : \left(\frac{d\Omega}{dt}\right)_0 = 0$$

8. On the mean motion: Corollary VI , Prop. *XXXXX: Corollary I-VI*

$$F_E = \mu\frac{r}{r^3}$$

9. On the formation of the inclination: Corollary X, Prop. *XXXXX: Corollary VII-XVII*

and

$$n \text{ at perhilion} < n \text{ at aphelion}$$

Proposition *LXXXXVI*. Problem *II*

We seek to determine the unequal motions of the satellites of Jupiter and Saturn from the motions of our Moon.

The afore mentioned elements that were originally conceived as variations, as now seen part of the whole motion of the Moon. These inequalities were found in other planet-satellite systems. Thus, from the mean annual regression of the nodes of 19.2862 in the Earth-Moon system, we can then infer that the mean regression to be found in the Jupiter-Callisto system. Then

$$\frac{\Delta\Omega_{\jmath}}{\Delta\Omega_{\oplus}} = \left(\frac{period_{\oplus}}{period_{\jmath}}\right)^2 \frac{period_{callisto}}{period_{Moon}} = \frac{1}{(11\cdot862)^2}\frac{16\cdot689}{27\cdot32}\times19\cdot286 = 8\cdot373 = 8\ 22'\ (1.679)$$

per hundred years,

in agreement with the value 8 24′.

As Newton pointed out the advance of the line of apsides can likewise be predicted. The observed motion in the Earth-Moon system is 'twice as fast' as originally predicted.

Propositions LXXXXVII-LXXXXVIII

Proposition *LXXXXVII*. Problem *III*

We seek to find the hourly increment of the area which the Moon, by a radius drawn to the Earth describes in a circular orbit.

Aside from the distinction between expressing the time-dependent values in terms of the nodical month and in the synodic month is now made, which will be better evaluated in section of the formation of the ascending node.

In comparing the changes made from Newton's numerical values with their analytical ones, such that

$$m^2 = (0 \cdot 0748025)^2 = 0 \cdot 005595414 \quad \frac{1000}{17818} \tag{1.680}$$

and

$$\frac{3}{2} m^2 = 0 \cdot 00839323 \quad \frac{100}{11915} \tag{1.681}$$

Thus by the 'constant of areas', where h at syzygies and quadratures, will be now replaced by

$$\frac{[h]}{[h_1]} = \frac{1 + \dfrac{3}{4} m^2}{1 - \dfrac{3}{4} m^2} = \frac{11965}{11865} \tag{1.682}$$

Since

$$m_{synodic} = m_{nodical} \frac{29 \cdot 531}{27 \cdot 322} = 1 \cdot 08085 m_{nodical} \tag{1.683}$$

equation (1.681) now expressed in terms of the nodical month becomes

$$\left(\frac{[h_1]}{[h_0]} \right)_{synodic-month} = \frac{11131}{11023} \tag{1.684}$$

Then in terms of the formation of the constant of areas, we will have

$$\frac{1}{2} d[h] = \frac{3}{4} F_T \left\langle r^2 \right\rangle_{AV} \sin 2\psi \, dt = \frac{1}{238 \cdot 3} \left\langle r^2 \right\rangle_{AV} \sin 2\psi \, dt$$

$$\frac{1}{2} d[h] = \frac{3}{4} m^2 \left\langle nr^2 \right\rangle_{AV} \sin 2\psi . ndt = \frac{1}{238 \cdot 3} \left\langle nr^2 \right\rangle_{AV} \sin 2\psi \left(n_{nodical} \right) dt \tag{1.685}$$

when as expressed in terms of nodical months, after multiplication by $1 \cdot 08085$, becomes

$$\frac{1}{2} d[h] = \frac{1}{220 \cdot 5} \left\langle nr^2 \right\rangle_{AV} \sin 2\psi \left(n_{nodical} \right) dt \tag{1.686}$$

Proposition *LXXXXVIII.* Problem *IV*

From the hourly motion of the Moon we seek to find its distance from the Earth.

The area which the Moon, by a given radius drawn to the Earth, does describes in every moment of time, is as the hourly motion of the Moon, and also the square of the distance of the Moon from the Earth conjointly. Therefore the distance of the Moon from the Earth varies directly as the square root of the area and inversely as the square root of the hourly motion, taken jointly.

Cor. *I.* Thus the diameter of the Moon is given; it is inversely as the distance of the Moon from the Earth.

Cor. *II.* Hence also then the orbit of the Moon can be more exactly defined.

Proposition *LXXXXIX.* Problem *V*

We seek to find the diameters of the orbit, in which the Moon would move, without the eccentricity.

In regard to this Proposition we look to further analyze the equations describing the Earth-Moon system in the section, 'the determination of *x*', of Chapter *11*. Most of the properties in this section have already been found in this area. One detail, though, was over looked. Newton had originally derived that this relation was based on the assumption that the orbit is a prolate ellipse.

The derivation of this relation is

$$r = 1 \pm x \cos 2\psi \tag{1.687}$$

Given that the orbit is considered to be an ellipse we will have

$$\frac{1}{r^2} = \frac{\cos^2 \psi}{a^2} + \frac{\sin^2 \psi}{b^2} : \left(a^2 < b^2\right) \tag{1.688}$$

We will have then successively

$$\frac{1}{r^2} = \frac{1}{2a^2b^2}\left[a^2\left(1-\cos 2\psi\right)+b^2\left(1+\cos 2\psi\right)\right]$$
$$= \frac{1}{2a^2b^2}\left[\left(a^2+b^2\right)+\left(b^2-a^2\right)\cos 2\psi\right]$$

(1.689)

or

$$\frac{1}{r} = \frac{\sqrt{\left(a^2+b^2\right)}}{ab\sqrt{2}}\left(1+\frac{b^2-a^2}{a^2+b^2}\cos 2\psi\right)^{1/2}$$

(1.690)

then using the following relations

$$a = 1-x$$
$$b = 1+x$$
$$a^2+b^2 = 2\left(1+x^2\right)$$
$$b^2-a^2 = 4x$$

(1.691)

we will have

$$r = 1 \pm x\cos N_\beta\psi + O\left(x^2\right)$$

(1.692)

Proposition *LL*. Problem *VI*

To find the actual motion of the Moon.

The elliptical orbit, described about the center, will not describe equal areas in equal time, due to the nature of its motion about the Earth. There will be a consistent loss and gain in longitude, such as a change in the angle ψ, relative to the hypothesized but unnatural circular motion. This inequality is natural.

The equations governing the elliptical orbit can be given as

$$x = r\sin\phi \qquad\qquad y = r\cos\phi \qquad\qquad \left(\phi = \frac{\pi}{2}-\psi\right)$$

(1.693)

and

$$\frac{x^2}{a^2} + \frac{y^2}{b^2} = 1 \tag{1.694}$$

where

$$\left(a < b\right) \tag{1.695}$$

also given as

$$\frac{1}{r^2} = \frac{\sin^2 \phi}{a^2} + \frac{\cos^2 \phi}{b^2} \tag{1.696}$$

This elliptical orbit can be defined, hypothetically as describing equal areas in equal times be defining the constant of areas h by the following equations

$$r^2\left(m\frac{d^2 r_1}{dt^2}(-)\sin\phi \oplus m\frac{d^2 r_2}{dt^2}\sin\phi \oplus m\frac{d^2 r_3}{dt^2}(-)\cos\phi \oplus m\frac{d^2 r_4}{dt^2}\cos\phi\right) = C \tag{1.697}$$

where

$$C = F_T + \frac{\mu}{r} \tag{1.698}$$

where h is defined as

$$h = \frac{C}{ab} \tag{1.699}$$

by differentiating the above equation with respect to time we then obtain

$$\sec^2\phi\left(m\frac{d^2 r_1}{dt^2}(-)\sin\phi \oplus m\frac{d^2 r_2}{dt^2}\sin\phi \oplus m\frac{d^2 r_3}{dt^2}(-)\cos\phi \oplus m\frac{d^2 r_4}{dt^2}\cos\phi\right) = \frac{a}{b}h\sec^2 ht \tag{1.700}$$

The left-hand side of this equation can be successively transformed with the aid of the above equations in the following manner

$$\frac{C}{r^2}\sec^2\phi = hab\sec^2\phi\left(\frac{\sin^2\phi}{a^2}+\frac{\cos^2\phi}{b^2}\right)=\frac{a}{b}h\left(1+\frac{b^2}{a^2}\tan^2\phi\right) \tag{1.701}$$

while the right-hand side of the equation becomes

$$\frac{a}{b}h\left(1+\tan^2 ht\right)=\frac{a}{b}h\left(1+\frac{b^2}{a^2}\tan^2 ht\right) \tag{1.702}$$

Now in the actual case we will have

$$h=C\times m \tag{1.703}$$

we assume then that

$$\tan\phi=\lambda\tan(h)t \tag{1.704}$$

where λ is a constant still to be determined. From before we moved forward to the following equation

$$\lambda(h)\sec^2 ht=\sec^2\phi\left(m\frac{d^2r_1}{dt^2}(-)\sin\phi\oplus m\frac{d^2r_2}{dt^2}\sin\phi\oplus m\frac{d^2r_3}{dt^2}(-)\cos\phi\oplus m\frac{d^2r_4}{dt^2}\cos\phi\right) \tag{1.705}$$

$$=\frac{C}{r^2}\sec^2\phi=C\sec^2\phi\left(\frac{\sin^2\phi}{a^2}+\frac{\cos^2\phi}{b^2}\right)=C\left(\frac{1}{b^2}+\frac{1}{a^2}\tan^2\phi\right)=C\left(\frac{1}{b^2}+\frac{\lambda}{a^2}\tan^2 ht\right)$$

therefore

$$C=\frac{\lambda(h)}{\left(\lambda^2/a^2\right)\sin^2(h)t+\left(1/b^2\right)\cos^2(h)t} \tag{1.706}$$

We now determine the value λ by determining the constant of areas C at quadratures and syzygies. At quadratures

$$\phi = ht = 0 \qquad \text{and} \qquad C = b^2 \lambda(h) = h_0 \qquad (1.707)$$

while at syzygies

$$\phi = ht = \pi/2 \qquad \text{and} \qquad C = a^2(h)/\gamma = h_1 \qquad (1.708)$$

then by combining the last two equations

$$\frac{h_0}{h_1} = \frac{b^2 \lambda^2}{a^2} \qquad (1.709)$$

we will then have

$$\lambda = \frac{a}{b}\sqrt{\frac{h_0}{h_1}} \qquad (1.710)$$

or, substituting for h_1/h_0 from equation (1.681), Prop. *LXXXXVII*, and remembering that $a/b = (1-x)/(1+x)$ we will have then

$$\lambda = \frac{a}{b}\left(1 - \frac{3}{4}\frac{P}{1 \pm P}\right) = \frac{1-x}{1+x}\left(1 - \frac{3}{4}\frac{P}{1 \pm P}\right) \qquad (1.711)$$

then substituting for x from equation (1.701), and for m its value $0 \cdot 074803$, we find then

$$\lambda = 0 \cdot 98126 \qquad (1.712)$$

The relating λ and the constant of areas h is then

$$\tan \phi = 0 \cdot 098126 \tan(h)t \qquad (1.713)$$

in agreement with

$$\frac{68 \cdot 6877}{70} = 0 \cdot 98125 \qquad (1.714)$$

at the octants

$$ht = \pi / 4, \qquad \tan\phi = 0 \cdot 098126, \qquad \phi = 44 \cdot 458 = 44\ 27'29'' \quad (1.715)$$

and

$$45 - \phi = 32'31'' \qquad (1.716)$$

according this exact calculation, the constant of areas is expressed in terms of nodical months; though it is better to express it in terms of synodic months. In the latter case it is better to take the value $32'31''$ and multiply it by 1.08085, giving us instead

$$45 - \phi = 35'9'' \qquad (1.717)$$

Then more generally equation (1.714) is written in the form

$$\tan\phi = (1-\varepsilon)\tan(h)t : (\varepsilon = 0 \cdot 01874) \qquad (1.718)$$

and then solving to the first order in ε by writing $\phi = ht + \delta$ for δ small, we find then

$$\delta = \pm\frac{1}{2}\varepsilon\sin(2ht) = \pm 0 \cdot 00937\sin 2\psi = \pm 32 \cdot 21''\sin 2\psi = \pm 34 \cdot 82''\sin 2\psi \qquad (1.719)$$

where ψ = radians and $\pm 32 \cdot 21''\sin 2\psi$ (h measured in sidereal time) and $\pm 34 \cdot 82''\sin 2\psi$ (h measured in synodic time)

This is the magnitude in terms of the mean distance of the Sun from the Earth, not accounting the difference arising from the curvature of the larger orbit, and the affect of the Sun's presence on the Moon in line with the presence of the Earth equally as well. The affect of the Sun's presence on the Moon given in terms as directly of the square of the ration of the time of the synodic revolution of the Moon (the time of the year being then given) and inversely as the cube of the ratio of the distance of the Sun from the Earth. Then the apogee of the Moon's orbit is given as $33'14''$ while the perigee is given a $37'11''$ as long as the eccentricity is to the transverse semidiameter of the larger orbit. It is given as $16\frac{15}{16}$ to 1000.

The formulation of the Moon's ascending node Ω: Proposition LLI

As we have shown earlier in, Proposition *XXXXVIII*, Corollary *XI*, the equations for the angle Ω in the Moon's orbit is given by

$$F_T = N_\alpha \left(F_E \oplus F_S\right)r = N_\alpha \left(F_E \oplus F_S\right)\left(x(t)e_x + y(t)e_y + z(t)e_z\right) \qquad (1.720)$$

Proposition *LLI*. Problem *VII*

We seek to find the hourly motion of the nodes of the Moon in a circular orbit.

As it has been shown in Proposition *XXXXVIII*, if we include the force due to the presence of the Sun, along with the Earth, the equations constitute the motion of the Moon, in terms of the ascending node in the plane of the ecliptic, determining the direction of the node Ω.

The force that directs the direction of the ascending node normal to vE, along the ecliptic is given as

$$F_\Omega = N_\alpha r \left(F_E \oplus F_S\right)r = N_\alpha r \left(F_E \oplus F_S\right)\left(x(t)e_x + y(t)e_y + z(t)e_z\right) \qquad (1.721)$$

Than making use of the relations given in Chapter 3, section 'accelerations and velocities along a curved orbit', we will have

$$F_\Omega = N_\alpha v \sin\iota \left(F_E \oplus F_S\right)r = N_\alpha v \sin\iota \left(F_E \oplus F_S\right)\left(x(t)e_x + y(t)e_y + z(t)e_z\right) \qquad (1.722)$$

and

$$\frac{r}{h} = \frac{r}{r^2 \left(m\frac{d^2 r_1}{dt^2}(-)\sin\phi \oplus m\frac{d^2 r_2}{dt^2}\sin\phi \oplus m\frac{d^2 r_3}{dt^2} - \cos\phi \oplus m\frac{d^2 r_4}{dt^2}\cos\phi \right)} \qquad (1.723)$$

$$= \frac{r}{rC} = \frac{1}{C} = \frac{1}{\omega_s}$$

where v_s denotes the tangential velocity along the ecliptic. We will have then

$$\frac{d\Omega}{dt} = N_\alpha r^2 \left(F_E \oplus F_S \right) r = N_\alpha r^2 \left(F_E \oplus F_S \right)\left(x(t)e_x + y(t)e_y + z(t)e_z \right) \qquad (1.724)$$

since

$$\sin v = \frac{PH}{PT} : \sin U = \frac{AZ}{AT} : \cos\psi = \frac{IT}{PT} \qquad (1.725)$$

the velocity of the nodes is as $IT \cdot PH \cdot AZ$ or as the product of the sines of the three angles $TPI, PTN,$ and STN.

We will conclude then

$$\langle d\Omega \rangle_{AV} = \pm\frac{3}{4}F_T \quad \text{per nodical month} \qquad (1.726)$$

we find then

$$\langle d\Omega \rangle_{AV} = \pm\left(\frac{90 \cdot 646'}{60m} \right) = \pm 20 \cdot 1967$$

$$\langle d\Omega \rangle_{AV} = \pm 20 \cdot 1967 \text{ per year} \qquad (1.727)$$

which equals $\pm 8 \cdot 2942''$ per hour.

We can then write

$$d\Omega = \pm 33 \cdot 1768'' = \left(\pm 33' 10^{th} 36^{IV}\right) R\left(+\cos U \cos \omega + \sin U \sin \omega \cos \imath\right) \qquad (1.728)$$

Then the hourly motion will be to $33' 10^{th} 33^{IV} 12^{V}$ as the product of the three angles $TPI, PTN,$ and STN (or of the distances of the Moon from the quadrature, of the Moon from the node, and of the node from the Sun) to the cube of the radius.

Corollary I

In this corollary the original equation for the ascending node may be given as

$$\frac{d\Omega}{dt} = \pm 33 \cdot 1768'' \frac{A = PDdM}{2A(S) = TPM} \frac{AZ^2}{AT^2} \text{ per hour} \qquad (1.729)$$

where ($A =$ area of) and $2A(S) = 2$, are the area of the sector.

Corollary II

At sygzygies we will have

$$\text{area of } PDdM = 2 \text{ area of sector } TMP$$

and

$$\left(\frac{d\Omega}{dt}\right)_{syzygies} = \pm 33 \cdot 1768'' \frac{AZ^2}{AT^2} = \pm 33 \cdot 1768'' \sin^2 U \text{ per hour} \qquad (1.730)$$

Therefore, averaged over all initial values of U

$$\left\langle \left(\frac{d\Omega}{dt}\right)_{syzygies}\right\rangle_{AV} = \pm 16 \cdot 588'' \text{ per hour} \qquad (1.731)$$

Proposition *LLII*. Problem *VIII*

We seek to find the hourly motions of the nodes of the Moon in an elliptical orbit.

In this proposition we consider here the generalization of Proposition *X* (where we found the time in which the body's place is found moving in a given parabola) in the case of the ordinary elliptic orbit.

As we have shown in Proposition *LXXXXIX*, we will have an prolate ellipse, with a ratio of semiaxis,

$$\frac{b}{a} = \frac{1-x}{1+x} : x = m^2\left(1 + \frac{19}{6}m\right) \tag{1.732}$$

From Prop. *LXXXXVIII*, the equation described about its center, we here question the regression of the nodes of this prolate ellipse.

Our original expression for Ω will be given as

$$\frac{d\Omega}{dt} = N_\alpha\left(F_E \oplus F_S\right)r = N_\alpha\left(F_E \oplus F_S\right)\left(x(t)e_x + y(t)e_y + z(t)e_z\right) \tag{1.733}$$

followed by

$$\frac{d\Omega}{dt} = N_\alpha\frac{1}{\upsilon_s}\left(F_S \oplus F_E\right)r = N_\alpha\frac{1}{\upsilon_s}\left(F_S \oplus F_E\right)\left(x(t)e_x + y(t)e_y + z(t)e_z\right)$$
$$= N_\alpha r^2\left(F_S \oplus F_E\right)r = N_\alpha r^2\left(F_S \oplus F_E\right)\left(x(t)e_x + y(t)e_y + z(t)e_z\right) \tag{1.734}$$

where it was earlier noted in equation (1.723), of Proposition *LXXXXVIII*, r and h are still constants.

Then since

$$F_E = \langle h\rangle_{av} / ab \tag{1.735}$$

Then we will have for the previous equation

$$\frac{d\Omega}{dt} = N_\alpha \frac{r^2}{ab/\langle h \rangle_{av}} (F_E \oplus F_S) r$$

$$= N_\alpha \frac{r^2}{ab/\langle h \rangle_{av}} (F_E \oplus F_S)\left(x(t)e_x + y(t)e_y + z(t)e_z\right)$$

(1.736)

Then using the following relations from equation (1.691) of Prop. *LXXXXIX* given as

$$r = 1 \pm x \cos N_\beta \psi$$

(1.737)

we will have then to $O(x^2)$

$$\frac{d\Omega}{dt} = N_\alpha \frac{\left(1 - N_\beta x \cos N_\beta \psi\right)^2}{ab/\langle h \rangle_{av}} (F_E \oplus F_S) r$$

$$= N_\alpha \frac{\left(1 - N_\beta x \cos N_\beta \psi\right)^2}{ab/\langle h \rangle_{av}} (F_E \oplus F_S)\left(x(t)e_x + y(t)e_y + z(t)e_z\right)$$

(1.738)

Corollary I

Then using these relations, then

$$\sin \omega = \frac{PH}{PT} : \sin U \frac{AZ}{AT} : PT = AT$$

(1.739)

and

$$PT = AT = r$$

(1.740)

we will then have

$$R = \left(\cos^2 U + \sin^2 U \cos \iota\right)$$

(1.741)

or given more precisely in terms of segment areas of the sections corresponding to the Moon's orbit, geometrically

$$R = \frac{1}{rPM} Kk.PD\left(\cos^2 U + \sin^2 U \cos \iota\right) \tag{1.742}$$

then given as

$$\text{areas of } PDdM \,/2 \text{ area of sector } TPM \left(\cos^2 U + \sin^2 U \cos \iota\right) \tag{1.743}$$

Corollary II

The velocity of the nodes is as $IT \cdot PH \cdot AZ$, or also as the product of the sine of the three angles TPI, PTN, and STN.

At syzygies

$$\text{area of } PDdM = 2 \text{ area of sector } TMP$$

and

$$\left(\frac{d\Omega}{dt}\right)_{syzgies} = \pm 33 \cdot 1768'' \left(\cos^2 U + \sin^2 U \cos \iota\right) \tag{1.744}$$

then when we average over all initial values of Ω, we will have

$$\left\langle \left(\frac{d\Omega}{dt}\right)_{syzgies} \right\rangle_{AV} = \pm 16 \cdot 588'' \text{ per hour} \tag{1.745}$$

We can now observe the fact that

$$\left\langle \left(\frac{d\Omega}{dt}\right)_{syzygies} \right\rangle_{AV} = \pm 16 \cdot 5884'' \times 0 \cdot 98165 = \pm 16 \cdot 284'' = \pm 16'' 16^{th} 38^{IV} \tag{1.746}$$

Using the following relation

$$\tan \chi = \frac{b^2}{a^2} \cot \theta \tag{1.747}$$

we will have then

$$\frac{d\Omega}{dt} = N_\alpha \frac{r^2}{ab[h]/\langle h \rangle_{av}} \left(\sin^2 \theta + \frac{b^2}{a^2} \cos^2 \theta \right) \frac{PDdM}{2TPM} (F_E \oplus F_S) r$$

$$= N_\alpha \frac{r^2}{ab[h]/\langle h \rangle_{av}} \left(\sin^2 \theta + \frac{b^2}{a^2} \cos^2 \theta \right) \frac{PDdM}{2TPM} (F_E \oplus F_S) \qquad (1.748)$$

$$(x(t)e_x + y(t)e_y + z(t)e_z)$$

At syzygies, where

$$r^2 = b^2 : \theta = \pi / 2 \qquad (1.749)$$

and

$$\text{area of } PDdm = 2 \text{ area of sector } TPM \qquad (1.750)$$

giving us

$$\left(\frac{d\Omega}{dt} \right)_{syzygies} = \pm 33 \cdot 1768'' \left[\frac{\langle h \rangle_{av}}{h} \right]_{syzygies} \frac{b}{a} \left(\cos^2 U + \sin^2 U \cos \iota \right) \qquad (1.751)$$

Then from Proposition *XXXXVIII*, we will have

$$\frac{b}{a} = \frac{1-x}{1+x} = 0 \cdot 98571 \quad \left(\frac{69}{70} \right) \qquad (1.752)$$

we can write then

$$\left(\frac{d\Omega}{dt} \right)_{syzygies} = \pm 33 \cdot 1768'' (1 - 0 \cdot 01429) \left[\frac{\langle h \rangle_{av}}{h} \right]_{syzygies} \left(\cos^2 U + \sin^2 U \cos \iota \right)$$

$$= \pm 32.703'' \left(= 32'42^{th}22^{IV} \right) \left[\frac{\langle h \rangle_{av}}{h} \right]_{syzygies} \left(\cos^2 U + \sin^2 U \cos \iota \right) \qquad (1.753)$$

Then for Newton's original value we will have

325

$$\left[\frac{\langle h\rangle_{av}}{h}\right]_{syzygies} = 1 - \frac{3}{4}\frac{F_T}{1\pm P} = (1 - 0\cdot 0045356)$$

$$\left[\frac{\langle h\rangle_{av}}{h}\right]_{syzygies} = 1\pm\frac{F_T}{h} = 1\pm 0\cdot 0045356 \tag{1.754}$$

hence then we will have

$$\left(\frac{d\Omega}{dt}\right)_{syzygies} = \pm 32\cdot 703'' \times 0\cdot 99546\left(\cos^2 U + \sin^2 U\cos\imath\right)$$

$$= \pm 32\cdot 5545''\left(\cos^2 U + \sin^2 U\cos\imath\right) = \pm 32''33^{th}16^{IV}\left(\cos^2 U + \sin^2 U\cos\imath\right) \tag{1.755}$$

This requirement follows from the following fact

$$(1 - 0\cdot 01429)(1 - 0\cdot 004536) \quad (1 - 0\cdot 01429 - 0\cdot 004536)$$

$$= (1 - 0\cdot 001883) \quad (1 - q) \tag{1.756}$$

(a) The equation for $d\Omega/dt$ for a Kepler ellipse

We will begin with the following equation

$$\frac{d\Omega}{dt} = N_\alpha\frac{r^2}{ab}\cdot(F_E\oplus F_S)r = N_\alpha\frac{r^2}{ab}\cdot(F_E\oplus F_S)\left(x(t)e_x + y(t)e_y + z(t)e_z\right) \tag{1.757}$$

where we have replaced $a^2\sqrt{(1-e^2)}$ by ab. Then using the following relation

$$(\pi ab)/\pi = h(p/2\pi) = h/F_T \tag{1.758}$$

where $p = $ period, then our original equation above will become

as similar to equation (1.723) of Proposition *LXXXXIX*. Then we will have

$$\frac{d\Omega}{dt} = N_\alpha \frac{r^2}{ab} \frac{\sin\psi}{abPL/MQ} \frac{A.PDdM}{A.S.TPM} \cdot (F_E \oplus F_S) r$$

$$= N_\alpha \frac{r^2}{ab} \frac{\sin\psi}{abPL/MQ} \frac{A.PDdM}{A.S.TPM} \cdot (F_E \oplus F_S)(x(t)e_x + y(t)e_y + z(t)e_z) \tag{1.759}$$

where $A. = $ area of, and $A.S. = $ area of sector. And $(\varphi = \pi/2 - \psi)$.

If χ denotes the angle in which the tangent at P makes with PT (like in the direction of r) we will have

$$PL = PM\cos(\chi - \varphi) \qquad \text{and} \qquad MQ = PM\sin\chi \tag{1.760}$$

and

$$PL/MQ = \cos(\chi - \varphi)/\sin\chi = (1 + \cot\varphi\cot\chi)\sin\varphi \tag{1.761}$$

then our original equation for $d\Omega/dt$ becomes

$$\frac{d\Omega}{dt} = N_\alpha \frac{r^2}{ab} \frac{1}{(1 + \cot\varphi\cot\chi)} \frac{A.PDdM}{A.S.TPM} \cdot (F_E \oplus F_S) r$$

$$= N_\alpha \frac{r^2}{ab} \frac{1}{(1 + \cot\varphi\cot\chi)} \frac{A.PDdM}{A.S.TPM} \cdot (F_E \oplus F_S)(x(t)e_x + y(t)e_y + z(t)e_z) \tag{1.762}$$

by then using the already known properties of conic sections and equations (20-24) from Chapter 11, section 65(e) we will have

$$\cot\varphi\cot\chi = \frac{e}{1 - e^2 a} \frac{r}{a}\cos\varphi = \frac{e\cos\varphi}{1 + e\cos\varphi} \tag{1.763}$$

which give us then for the prior equation (1.762)

$$\frac{d\Omega}{dt} = N_\alpha \frac{r^2}{ab} \frac{1}{(1 + e\cos\varphi/(1 + e\cos\varphi))} \frac{A.PDdM}{2A.S.TPM} \cdot (F_E \oplus F_S) r$$

$$= N_\alpha \frac{r^2}{ab} \frac{1}{(1 + e\cos\varphi/(1 + e\cos\varphi))} \frac{A.PDdM}{A.S.TPM} \cdot (F_E \oplus F_S)$$

$$(x(t)e_x + y(t)e_y + z(t)e_z) \tag{1.764}$$

At syzygies

$$\varphi = \pi/2 : r = l = b^2/a \quad \text{and} \quad \text{area of } PDdM = 2 \text{ area of sector } TPM \quad (1.765)$$

Propositions LLIII and LLIV

Proposition LLIII. Problem VIV

In Propositions LXXXXIX and LL we have shown that

$$\left(\frac{d\Omega}{dt}\right)_{syzygies} = \pm R\left(\cos^2 U + \sin^2 U \cos \iota\right) \text{per sidereal year} \quad (1.766)$$

where

$$R = 20 \cdot 1967 \quad \text{for a circular orbit} \quad (1.767)$$

but for an elliptical orbit we will have

$$R = 20 \cdot 1967 \left(1 - q\right) = 19 \cdot 826 \left(= 19\,49'34''\right) \quad (1.768)$$

described about the center.

Here Newton had considered that R is as the mean regression of the nodes.

But it seems that the if the node were drawn back, after every hour, to its original position after a complete revolution, the Sun would, after the following year, be back in the same node as the year before.

Following then for an assigned value for U, the Sun in the course of a year travels 360 degrees, and the node in the same time would be carried $2R\left(\cos^2 U + \sin^2 U \cos \iota\right)$, the mean motion for an increment of dU in U is given then as

$$\pm \frac{2R\left(\cos^2 U + \sin^2 U \cos \iota\right)}{360 + 2R\left(\cos^2 U + \sin^2 U \cos \iota\right)} dU = \pm \frac{Q\left(\cos^2 U + \sin^2 U \cos \iota\right)}{1 + Q\left(\cos^2 U + \sin^2 U \cos \iota\right)} du \quad (1.769)$$

where

$$Q = R/180 \qquad (1.770)$$

Then the mean regression coefficient, as the Sun continues on its onward course in its annual orbit, is given as

$$\bar{Q} = \int_0^{\pi/2} \frac{Q\left(\cos^2 U + \sin^2 U \cos \iota\right)}{1 + Q\left(\cos^2 U + \sin^2 U \cos \iota\right)} dU \qquad (1.771)$$

then evaluating this integral we will have then

$$\bar{Q} = \frac{1}{2}\pi\left[1 - \frac{1}{\sqrt{(Q+1)}}\right] \qquad (1.772)$$

For Newton's original value of $R = 19\cdot826$ we will have $Q = 0\cdot11014$ and then

$$\bar{Q} = \frac{1}{2}\pi(1 - 0\cdot949094) = \frac{1}{2}\pi \times 0\cdot050904 = 0\cdot07996 \qquad (1.773)$$

which can be contrasted with

$$\bar{Q} = \frac{60}{793} = 0\cdot0756 \qquad (1.774)$$

where Newton had found this by the method of infinite series, almost.

During a sidereal year the Sun is carried by the following amount

$$R(1 - \bar{Q}) = 19\cdot826 \times 0\cdot92004 = 18\cdot2407 = 18\,14'15'' \qquad (1.775)$$

thus the regression coefficient is given then as

$$\pm\frac{18\cdot2407 \times 360}{360 - 18\cdot2407} = \pm\frac{18\cdot2407 \times 360}{341\cdot7593} = \pm19\cdot214 = \pm19\,12'50'' \qquad (1.776)$$

329

Proposition *LLIV*. Problem *X*

We seek to find the nodes of the Moon.

To figure the true motions of the nodes, we must integrate the original equation for $d\Omega/dt$, from Prop. *LXXXXIX* for a circular orbit and from the original equation in Prop. *LL*, for the elliptical orbit.

As we have shown from Prop. *XXXXVIII*: Corollaries *I-VI*, for a circular orbit, can thus be integrated with the following substitutions

$$\psi = (1 \pm F_T)v : U = m\omega : d\omega = F_T dt \tag{1.777}$$

where retaining only the dominant term among the periodic terms we will have

$$\Omega = \left(\pm \frac{3}{4} F_T nt + \frac{3}{8} P \sin 2U \right) \tag{1.778}$$

Then the first term in the following equation gives a secular decrease in Ω, as given by the mean regression coefficient R, while the second term represents a semi-annual periodic variation with an amplitude of

$$\frac{3}{8} P = 1 \cdot 6072 \tag{1.779}$$

For a semi-synodic-year the following amplitude is

$$1 \cdot 6072 \ / 1 \cdot 08085 = 1 \cdot 4887 \ = 1\,29'20'' \tag{1.780}$$

Then from Prop. *LXXXXIX*,

$$\Omega = \left(\pm \frac{3}{4} F_T nt + \frac{3}{8} P \sin 2U \right)\left(1 - \frac{1}{2} q \right) \tag{1.781}$$

The secular regression coefficient is now given as

$$R = \frac{1}{2}\left(20 \cdot 197 \ + 19 \cdot 826 \ \right) = 20 \cdot 012 \ = 20\,0'39'' \text{ per year} \tag{1.782}$$

also given as

$$R = 16 \cdot 436'' = 16'26^{th}10^{IV} \quad \text{per hour} \tag{1.783}$$

Then for the amplitude of the semi-annual periodic form for the elliptical orbit, we will have

$$\frac{3}{8}P\left(1-\frac{1}{2}q\right)/1\cdot08085 = 1\cdot4733 = 1\,28'24'' \tag{1.784}$$

The inclination in terms of the newly developed Lunar theory

In Propositions *LLV* and *LLVI* we focus on the development of the inclination, a topic we have discussed earlier in Prop. *XXXXVIII*: Corollaries X ,

$$\begin{aligned}
\frac{d\iota}{dt} &= N_\beta\left(r\cos v\right)\left(F_E \oplus F_S\right)r \\
&= N_\beta\left(r\cos v\right)\left(F_E \oplus F_S\right)\left(x(t)e_x + y(t)e_y + z(t)e_z\right) \\
&= N_\beta \frac{r^2}{a^2\sqrt{(1-e^2)}}\left(F_E \oplus F_S\right)r \\
&= N_\beta \frac{r^2}{a^2\sqrt{(1-e^2)}}\left(F_E \oplus F_S\right)\left(x(t)e_x + y(t)e_y + z(t)e_z\right)
\end{aligned} \tag{1.785}$$

Proposition *LLV*. Problem *XI*

To find the hourly form of the inclination of the Moon's orbit to the plane of the ecliptic.

The total force that determines the nature of the Moon's orbital inclination is given then as

$$\frac{d\iota}{dt} = N_\beta (r \cos v)(F_E \oplus F_S)(x(t)e_x + y(t)e_y + z(t)e_z)$$

$$= N_\beta \frac{r^2}{a^2 \sqrt{(1-e^2)}}(F_E \oplus F_S)(x(t)e_x + y(t)e_y + z(t)e_z) \qquad (1.786)$$

Then by making use of the elementary relations given earlier in Chapter 4, section 20, we will have then

$$\frac{d\iota}{dt} = \frac{\upsilon_s}{\rho} \qquad (1.787)$$

and

$$F_T = \frac{\upsilon_s^2}{\rho} = \frac{\omega}{\rho}\frac{h}{r} \qquad (1.788)$$

where ρ denotes the radius of curvature of the orbit and υ_s is the tangential velocity along it at P.

We will then have

$$\frac{d\iota}{dt} = N_\beta (r^2 \cos v)(F_E \oplus F_S)(x(t)e_x + y(t)e_y + z(t)e_z) \qquad (1.789)$$

Alternatively, by comparison with equation (1.728), Prop. *LLI* we can also write

$$\frac{d\iota}{dt} = \pm 33'10'''36^{IV} \cos vR \text{ per hour} \qquad (1.790)$$

Now by looking at the accompanying figure (which is the same as the figure in the context of Proposition *LLI*, with the additions in which the neighboring lines KP and kM are then extended to meet TF, normal to nN, at D and d), we observe then

$$\cos \psi = \frac{IT}{PT} : \sin \frac{AZ}{AT} : \cos v = \frac{TG}{PT} \qquad (1.791)$$

we will have then

$$\frac{d\iota}{dt} = \pm 33''10'''36^{IV}\frac{TG}{PT}\left(\frac{IT}{AT}\frac{AZ}{AT}\sin\iota\right) \text{ per hour} \tag{1.792}$$

Then the inclination will be to the angle $33''10'''33^{IV}$ as $IT.AZ.TG.\dfrac{Pp}{PG}$ to AT^3

Pp is to PG as the sine of the inclination to the radius.

Corollaries I – III

Here, referring to the figure again, we will have

$$\cos\psi = \frac{Kk}{PM}:\cos v\,\frac{TG}{PT}:TG = PQ = PD\cos U \tag{1.793}$$

we can then write equation (1.788) as

$$\frac{d\iota}{dt} = N_\alpha\frac{A.PDdM}{2A.S.TPM}\left(F_E \oplus F_S\right)\left(x(t)e_x + y(t)e_y + z(t)e_z\right) \tag{1.794}$$

Corollary IV

When the node is at quadrature

$$U = \pi/2 \qquad \text{and} \qquad \psi = v \pm \pi/2 \tag{1.795}$$

we will then have

$$\frac{d\iota}{dt} = +33\cdot1768''rR \quad \text{per hour} \tag{1.796}$$

In the time in which the Moon describes its orbit from the syzygies to the quadratures, namely

$$\frac{1}{4}\text{ period }=\frac{\pi}{2n}=177\cdot167\quad\text{hours}\qquad(1.797)$$

the inclination is then given as

$$\frac{d\iota}{dt}=\frac{1}{\pi}\left(177\cdot167\times33\cdot1768''\right)rR\quad\text{degrees}\qquad(1.798)$$

where

$$U\to\psi\qquad(1.799)$$

then we will have

$$\frac{d\iota}{dt}=163\cdot612''\left(=2'43''37'''\right)R\qquad(1.800)$$

both of which we will have in terms of $\frac{1}{4}$ period.

Proposition *LLVI*. Problem *XII*

At a given time, we find the inclination of the Moon's orbit to the plane of the ecliptic.

We start by replacing U,ψ and v by

$$U=F_T(v):\psi=\left(1\pm FT\right)v\quad\text{and}\quad dv=F_T dt\qquad(1.801)$$

we will have then

$$\iota-\iota_0=8\cdot4326'\cos 2U-1\cdot2616'\sin\left(2\psi+U\right)\sin U\qquad(1.802)$$

From this latter equation we will have the following table of values

$$(i): U = 0 : (\iota - \iota_0) = +8 \cdot 4326'$$
$$(ii): U = \pi / 4 : (\iota - \iota_0) = -0 \cdot 6303' (\sin 2\psi + \cos 2\psi)$$
$$(iii): U = \pi / 2 : (\iota - \iota_0) = -8 \cdot 4326' - 1 \cdot 2616' \cos 2\psi$$
$$(iv): \psi = 0 : (\iota - \iota_0) = +8 \cdot 4326' \cos 2U - 1 \cdot 2616' \sin^2 U$$
$$(v): \psi = 0 : (\iota - \iota_0) = +8 \cdot 4326' \cos 2U - 1 \cdot 2616' \sin U \cos U$$
$$(vi): \psi = 0 : (\iota - \iota_0) = +8 \cdot 4326' \cos 2U - 1 \cdot 2616' \sin^2 U$$

$$(1.803)$$

The maximum of the extreme tendencies of the orbital inclination is given as $16 \cdot 865' (= 16'52''1''')$

The difference in inclination between quadratures and syzygies of the Moon is $2.523 \sin^2 U$ or $(= 2'31'')$ for $U = \pi / 2$.

When the nodes are in the syzygies, the inclination of the Moon, according to its positions, stays the same, while if the nodes are in the quadratures the inclination is less when the Moon in the syzygies then when it is in the quadratures, by an estimated value of $2'43''$.

Finally from equations (i) and (iv) we will have for

$$\psi = 0 \quad \text{when} \quad U = 0 \quad \text{and} \quad U = \pi / 2 \qquad (1.804)$$

we will have respectfully

$$\iota - \iota_0 = 8 \cdot 4326' \quad \text{and} \quad \iota - \iota_0 = -8 \cdot 4326' - 1 \cdot 2616' = -9 \cdot 6942' \qquad (1.805)$$

there difference is

$$17 \cdot 1268' \qquad (1.806)$$

and from equations (i) and (iii) it follows then that

$$\psi = \pi / 2 \quad \text{when} \quad U = 0 \quad \text{and} \quad U = \pi / 2 \qquad (1.807)$$

we will then have respectfully

$$\iota - \iota_0 = 8 \cdot 4326' \quad \text{and} \quad \iota - \iota_0 = -8 \cdot 4326' + 1 \cdot 2616' = -7 \cdot 171' \qquad (1.808)$$

335

and their difference is

$$15 \cdot 6036'$$ (1.809)

The seemingly normal orbit, similar to the Earth's orbit, we will have an inclination of 5 1′. Here though the inclination of the Moon at the syzygies will be 5 19′ and at the quadratures it will be 5 4′.

Scholium

As we have discussed, in terms of the lunar problem in Corollaries *VI* and *VII* of Propositions *LXVI* of Book *I*, we will look at the problem of the annual equation, including the motion of the apogee.

(b) The annual equation

In Corollary *VI* of Proposition *XXXXVIII*, the ordinary period of the orbit of the Moon is given as

$$P = 2\pi \left(N_\alpha \left(\frac{a}{R} \right)^3 (F_E \oplus F_S) \theta_r \right) \left(x(t)e_x + y(t)e_y + z(t)e_z \right)$$ (1.810)

where a_M and a denote the semimajor axis of the orbits of the Moon and of the Earth. Given the equation for the hypothetical circular orbit of the Moon, the mean period is then given as

$$\bar{P} = P_M \left(1 + (F_T) \right)$$ (1.811)

where P_M denotes the hypothetical period of the Moon. If we include the presence of the Sun, on the normal period of the Moon, the change is given by the following value

$$\Delta \bar{P} = 0 \cdot 9914 \quad \text{hours}$$ (1.812)

Now returning to the original equation for the annual motion, the distance between the Earth and the Sun, R , must be included in the formulating the mean period of the Moon. To show this we will replace $(a\ /R\ )$, with the first order in the eccentricity of the Earth's orbit, by

$$\frac{a}{R} = 1 + e\ \cos\varphi \qquad (1.813)$$

we will then have for the annual equation

$$P = 2\pi\left(N_\alpha\left(1+e\ \cos\varphi\right)^3\left(F_E \oplus F_S\right)\theta_r\right)\left(x(t)e_x + y(t)e_y + z(t)e_z\right) \qquad (1.814)$$

or, averaging over ψ , we will have

$$\bar{P} = P_M\left(1+\left(F_T\right)\left(1+e\ \cos\varphi\right)\right) \qquad (1.815)$$

Therefore

$$\bar{P}_{,\varphi} = \pm P_M\left(F_T\right)e\ \sin\varphi \qquad (1.816)$$

Since the orbital period of the Earth is given as

$$P\ = P_M$$

The annual equation for the Moon can be written then as

$$\text{annual equation} = \pm F_T e\ \sin\varphi \qquad (1.817)$$

For e we will have the currently used value of $0\cdot016709$, giving us then

$$\text{annual equation} = -12\cdot92'\sin\varphi = -\left(12'56''\right)\sin\varphi \qquad (1.818)$$

Due to both the presence of the Earth and the Sun, the annual equation of the mean motion of the Moon from the dilatation which the Moon is affected by, is primarely due to the presence of the Sun. The force depends on both the perigean Sun and the apogean Sun, the first being greater and the latter being less. The Moon moves slower in the dilated orbit and faster in the contracted orbit. The annual equation vanishes in the apogee and perigee of the Sun. The mean distance of the Sun from the Earth rises to be about $11'50''$: in other distances from the Sun it is directly proportional to the equation of the Sun's center, and is added to the mean motion of the moon, while the Earth is passing from its perihelion to its aphelion, and subtracted while the Earth is in the opposite semicircle. The radius of the great orbit 1000, and $16\frac{7}{8}$ for the Earth's eccentricity, this equation, at the largest magnitude being $11'49''$.

(c) The motion of the apogee and the 'Portsmouth equation'

In the preceding propositions we have successfully evaluated the equations for the orbit, the period and the annual equation (all requiring a new approximation). There were also given first-order equations for the regression of the nodes, and the eccentricity and the inclination of the orbit.

In previous propositions we have considered the eccentricity of the lunar orbit due to the presence of the Sun. But in fact the orbital motion of all moons and related satellites must include the presence of the Sun in a ratio equal to the presence of the Earth or other planetary bodies.

One of the more interesting considerations involves the motion of the lunar apogee. The eccentricity of the lunar orbit in an elliptic orbit of relatively small eccentricity. Both the presence of the Earth and the Sun imply forces that work in both radial and parallel, but opposite radial directional forces.

The hourly motion of the apogee can be represented by the following formula

$$\frac{d\varpi}{dt} = \frac{\mu}{r^3}r + F\left(\frac{11}{2}\cos\left(2v' \pm 2\varpi\right)\right) \qquad (1.819)$$

The same formula, restricted to its first few terms, is then given as

$$\frac{d\varpi}{dt} = \frac{\mu}{r^3}r + F\left(5\cos\left(2v' \pm 2\varpi\right)\right) \tag{1.820}$$

Aside from replacing $11/2$ by 5, is the same formula as (1.819). The mean annual motion of the apogee is $40\ 41'5''$.

Consider the more general equation

$$\frac{d\varpi}{dt} = F_T\left[1 + \alpha\cos\left(2F_Tv \pm 2\varpi\right)\right] \tag{1.821}$$

where α is a constant, not specified for right now. While v' in equations (1.818) and (1.819) has been replaced by F_Tv , we will then have

$$v' = v - \psi = v - \left(1 \pm F_T\right)v = mF_T \tag{1.822}$$

With the following substitution

$$\varpi = F_Tv + \phi \tag{1.823}$$

then equation (1.819) will become

$$\frac{d\phi}{dv} = F_T\alpha\cos\left(2qv \pm 2\phi\right) \tag{1.824}$$

where

$$q = P\left(1 + F\ \right) \tag{1.825}$$

By another substitution we will have

$$\phi = -\frac{1}{2}y + qv \tag{1.826}$$

equation (1.823) can then be written as

$$\frac{dy}{dv} = a\cos y + b \tag{1.827}$$

where

$$a = -\frac{3}{2}F_T\alpha \quad \text{and} \quad b = 2q \tag{1.828}$$

the solution of equation (1.826) is then given as

$$\tan^{-1}\left[\left(\frac{b+a}{b-a}\right)^{1/2}\cot\frac{y}{2}\right] + \frac{1}{2}v\left(b^2 \pm a^2\right)^{1/2} = C \tag{1.829}$$

where C is a constant of integration. For a and b given in equation (1.828) the solution to this can be given as

$$\left(\frac{1-\varepsilon}{1+\varepsilon}\right)^{1/2}\cot\frac{y}{2} = \tan\left[C \pm qv\left(1-\varepsilon^2\right)^{1/2}\right] \tag{1.830}$$

where

$$\varepsilon = \frac{\alpha P}{\left(1 \pm 3P/4\right)} \tag{1.831}$$

With the requirement

$$\phi = 0 \quad \text{when} \quad \alpha = 0 \tag{1.832}$$

determines

$$C = \frac{1}{2}\pi \tag{1.833}$$

then the solution reduces to

$$\tan\frac{1}{2}y = \left(\frac{1-\varepsilon}{1+\varepsilon}\right)^{1/2}\tan\left[qv\left(1-\varepsilon^2\right)^{1/2}\right] \tag{1.834}$$

Finally by combining equations (1.826) and (1.834) we will have

$$\phi = qv \pm \tan^{-1}\left\{\left(\frac{1-\varepsilon}{1+\varepsilon}\right)^{1/2} \tan\left[qv\left(1-\varepsilon^2\right)^{1/2}\right]\right\} \tag{1.835}$$

For $v = \pi$ and $\alpha = \dfrac{11}{2}$ we find from a numerical evaluation of equation (1.835) as

$$\phi = 0 \cdot 2218 - 0 \cdot 1504 = 0 \cdot 0714 \tag{1.836}$$

We conclude then that α is numerically too large by a factor of six, to account for a discrepancy of a factor of 2.

An estimate of α that can account for the factor 2 can be made by solving equation (1.835) to $O(\varepsilon)$ assuming that ε is small.

Letting

$$qv = \frac{1}{2}y + \phi \tag{1.837}$$

we find then that to $O(\varepsilon)$ the solution of equation (1.834) is

$$\phi = \frac{1}{2}\varepsilon \sin(2\sin qv) \tag{1.838}$$

or in our original scheme of approximation

$$\delta = \varepsilon qv = F_T \alpha v \tag{1.839}$$

or in other words

$$\varpi = F_T(1+\alpha)v \tag{1.840}$$

we conclude that

$$\alpha \leq 1 \tag{1.841}$$

18

The precession of the equinoxes

Introduction

As produced in Lemmas I, II and III, we have considered the notion of dynamics of rigid bodies: the moment of inertia (in relation to the wheel with a given circular motion about a center); the moment of momentum (which measures the motions of the whole body about its axis or rotation) and the torque (the power to wheel about), exerted by an previously defined force on a non-spherical body; and the underlying principles of gyroscopic motion. All of these notions are defined in the context of a dynamical theory of the precession of the equinoxes which derives from the given shape, directly, of the Earth and the inclination of its axis of rotation to the plane of the ecliptic.

On the precession of the equinoxes: a current treatment

We list here a set of definitions relating a series of specific notations

Frame of reference: (x, y, z)

Unit vectors along x, y and z:

$(i, j, k), r = xi + yj + zk$

Force: $F = Xi + Yj + Zk$

Equation of motion: $F = mr$

Moment of force: $M = r \times F$

Angular velocity: $\qquad\qquad\qquad\qquad\qquad\qquad \omega = \omega_x i + \omega_y j + \omega_z k$

Velocity associated with ω: $\qquad\qquad\qquad\qquad\qquad v = \omega \times r$

Angular momentum of an element of mass, dm: $\qquad\qquad dh = dm \times \omega$

Combining the definitions for dh and v we will have

$$dh = dmr \times v = dmr \times (\omega \times r) = dm\left[r^2\omega - r(r \cdot \omega)\right] \qquad (1.842)$$

Then the angular momentum of a rigid body about the common center of rotation is given by

$$H = \int_V dh = \omega \int_V r^2 dm - \int_V r(r \cdot \omega) dm \qquad (1.843)$$

here the integration is over the volume V, occupied by the body in question. Then reducing the expression for H, we will have

$$
\begin{aligned}
H = &\int_V dm\left[(x^2 + y^2 + z^2)(\omega_x i + \omega_y j + \omega_z k) - (x\omega_x + y\omega_y + z\omega_z)(xi + yj + zk)\right] \\
= &\int_V dm\left\{\left[\omega_x(x^2 + y^2 + z^2) - (\omega_x x^2 + \omega_y xy + \omega_z xz)\right]\right\} i \\
&+ \left[\omega_x(x^2 + y^2 + z^2) - (\omega_x zx + \omega_y y^2 + \omega_z zy)\right] \\
&+ \left[\omega_x(x^2 + y^2 + z^2) - (\omega_x xz + \omega_y yz + \omega_z z^2)\right]
\end{aligned}
\qquad (1.844)
$$

The components of H are given as

$$
\begin{aligned}
H_x &= +A\omega_x - F\omega_y - E\omega_z \\
H_y &= -F\omega_x + B\omega_y - D\omega_z \\
H_z &= -E\omega_x - D\omega_y + C\omega_z
\end{aligned}
\qquad (1.845)
$$

where

$$\left\{A = \int_V dm(y^2 + z^2), \int_V dm(z^2 + z^2), \int_V dm(x^2 + y^2)\right\} \qquad (1.846)$$

and

$$D = \int_V dm yz, \qquad E = \int_V dm zx, \qquad F = \int_V dm xy \qquad (1.847)$$

define the moment of inertia tensor. Then if the coordinate frame is chosen along the principle axes of the momentum of inertia tensor, then

$$D = E = F = 0 \qquad (1.848)$$

and

$$\left\{ A = \int_V dm\left(y^2 + z^2\right), \int_V dm\left(z^2 + z^2\right), \int_V dm\left(x^2 + y^2\right) \right\} \qquad (1.849)$$

and the components of H are

$$H_x = A\omega_x, \qquad H_y = B\omega_y, \qquad H_z = A\omega_z \qquad (1.850)$$

In particular, if the body is axisymmetric about the axis or rotation in the direction k, then

$$A = B \neq C \qquad (1.851)$$

Finally the equation of motion governing H is

$$\frac{dH}{dt} = \frac{d}{dt}\int_V dm\, r \times v = \int_V dm\left(\frac{dr}{dt} \times v + r \times \frac{dv}{dt}\right) = \int_V r \times dF = \int_V dm = M \qquad (1.852)$$

(a) Euler's equations

We consider now a given rigid body with a given angular velocity, ω, about a center fixed in the body. Let $OXYZ$ be a coordinate frame attached rigidly to the body and also sharing its motion. Then to describe the motion of the body, we must choose another frame $OX_1Y_1Z_1$ fixed in space with the same center.

Let (i,j,k) denote unit vectors along OX, OY and OZ; and let ϖ be the angular velocity of the frame $OXYZ$ (i.e. of the center of inertia of the body) and H then angular momentum of the body in the fixed frame $OX_1Y_1Z_1$. If $(\omega_x, \omega_y, \omega_z)$ and (H_x, H_y, H_z) denote the components of ω and H resolved along the axes OX, OY and OZ fixed in the body, then the given definitions

$$\omega = \omega_x i + \omega_y j + \omega_z k \qquad \text{and} \qquad H = H_x i + H_y j + H_z k \qquad (1.853)$$

hold. Since i, j and k are no longer time-independent (as viewed in the frame $OX_1Y_1Z_1$, fixed in space) we will allow it when obtaining the equations of motion of H. Then

$$\frac{dH}{dt} = i\frac{dH_x}{dt} + j\frac{dH_y}{dt} + k\frac{dH_z}{dt} + H_x\frac{di}{dt} + H_y\frac{dj}{dt} + H_z\frac{dk}{dt} \qquad (1.854)$$

To evaluate the time derivative of i, j and k we consider them as being position vectors, r drawn from O. And, since for any vector r,

$$\frac{dr}{dt} = \omega \times r = \begin{vmatrix} i & j & k \\ \omega_x & \omega_y & \omega_z \\ r_x & r_y & r_z \end{vmatrix} \qquad (1.855)$$

we can then replace r in this equation by i, j and k. Therefore,

$$\frac{di}{dt} = \begin{vmatrix} i & j & k \\ \omega_x & \omega_y & \omega_z \\ 1 & 0 & 0 \end{vmatrix} = \omega_z j - \omega_y k \qquad (1.856)$$

and the expressions for $\dfrac{di}{dt}$ and $\dfrac{dk}{dt}$ can be obtained by cyclically permuting (i,j,k) and $(\omega_x, \omega_y, \omega_z)$. Inserting the resulting expression in equation (1.854), we will have then

$$\frac{dH}{dt} = iH_x + jH_y + kH_z + H_x\left(\omega_z j - \omega_y k\right) + H_y\left(\omega_x k - \omega_z i\right) + H_z\left(\omega_y i - \omega_x j\right)$$

$$= i\left(H_x + \omega_y H_z - \omega_y H_z\right) + j\left(H_y + \omega_z H_x - \omega_x H_z\right) + k\left(H_z + \omega_x H_y - \omega_y H_x\right) \qquad (1.857)$$

$$= iM_x + jM_y + kM_z$$

If the unit vectors i, j and k are chosen along the principal axes of the inertia tensor, equation (1.850) holds and equation (1.857) gives us

$$A\omega_x + (C-B)\omega_z\omega_y = M_x$$
$$B\omega_x + (A-C)\omega_z\omega_x = M_y \qquad (1.858)$$
$$C\omega_x + (B-A)\omega_x\omega_y = M_z$$

and if the body is axissymmetric about the rotation axis in the direction k, when $A = B$, equation (1.858) reduces to

$$A\omega_x + (C-A)\omega_y\omega_z = M_x$$
$$A\omega_y + (A-C)\omega_z\omega_x = M_y \qquad (1.859)$$
$$C\omega_z = M_z$$

(b) Euler's angles

The definition of Euler's angles, $\theta, \phi,$ and ψ, and the equations of motion governing them are shown in the accompanying diagrams. We have

$$\omega_x = \theta\sin\phi - \psi\sin\theta\cos\phi$$
$$\omega_y = \theta\cos\phi + \psi\sin\theta\sin\phi \qquad (1.860)$$
$$\omega_z = \phi + \psi\cos\theta$$

where ψ and θ determine the rates or precession and nutation, respectfully.

(c) The equations governing precession and nutation

Before we can correctly apply equations (1.859) and (1.860) to the problem of precession and nutation, we must first determine the moment of force, derived from the presence of the Sun, or more generally from a distant body, denoted as S. We can determine the required moment M in two steps: first in the frame $OXYZ$, rigidly attached to the center of the Earth; and then transform it to the frame in which the center of the Earth describes its orbit about S in the plane of the ecliptic.

The given configuration in the frame $OXYZ$ is illustrated in the adjoining diagram where:

S is the position of the distant body;

O the center of the Earth; and

P is the location of an element of mass of the Earth at

$$\rho = ix + jy + kz$$
$$R = OS, \quad r = PS, \quad \theta = \angle\left(OS, j\right) \tag{1.861}$$

With these definitions,

$$R = OQ + QS = jR\cos\theta - kR\sin\theta \tag{1.862}$$

and

$$r = R - \rho = -ix + j\left(R\cos\theta - y\right) - k\left(R\sin\theta + z\right) \tag{1.863}$$

and then the acting force on dm is

$$dF = F_{uv(\alpha)}dm\frac{r}{r^3} \tag{1.864}$$

where $\left(r = |r|\right)$ and M is the mass of the Sun. The moment of this force dF is

$$dM = \rho \times dF = \frac{F_{uv_{(\alpha)}}M}{r^3}dm \begin{vmatrix} i & j & k \\ x & y_\alpha & y_\beta \\ (-xR\sin\theta) & (-y_\alpha(R\cos\theta+\sin\theta)) & (-y_\beta(R\cos\theta+\sin\theta)) \end{vmatrix} \quad (1.865)$$

$$= \frac{F_{uv_{(\alpha)}}M}{r^3}dm\left[-iR(x\sin\theta)+(-)jR(\cos\theta+\sin\theta)+(-)kR(\cos\theta+\sin\theta)\right]$$

Therefore

$$dM_x = \pm\frac{F_{uv_{(\alpha)}}Mdm}{r^3}Rx\sin\theta$$

$$dM_y = \pm\frac{F_{uv_{(\alpha)}}Mdm}{r^3}Ry_\alpha(\sin\theta+\cos\theta) \quad (1.866)$$

$$dM_z = \pm\frac{F_{uv_{(\alpha)}}Mdm}{r^3}Ry_\beta(\sin\theta-\cos\theta)$$

Next we must express r in terms of ρ and R with the aid of the following relation,

$$r^2 = R^2 + \rho^2 - 2R\cdot\rho$$
$$= R^2 + \rho^2 - 2R\left(j(\sin\theta+\cos\theta)-k(\sin\theta+\cos\theta)\right)\cdot\left(ix+jy_\alpha+ky_\beta\right) \quad (1.867)$$

or also as

$$r^2 = R^2\left[1+\left(\frac{\rho}{R}\right)^2-\frac{2}{R}\left(j(\sin\theta+\cos\theta)-k(\sin\theta-\cos\theta)\right)\cdot\left(ix+jy_\alpha+ky_\beta\right)\right] \quad (1.868)$$

and then

$$\frac{1}{r^3} \quad \frac{1}{R^3}\left[1+\frac{3}{R}\left(j(\sin\theta+\cos\theta)-k(\sin\theta-\cos\theta)\cdot\left(ix+jy_\alpha+ky_\beta\right)\right)\right] \quad (1.869)$$

Inserting this last expression in equation (1.866) we will have

$$M_T = \left(\frac{F_{uv_x} M_x}{R^2} \int_{V_x} (Rx\sin\theta) \left[1 + \frac{3}{R} \left(\frac{j(\sin\theta + \cos\theta) - k(\sin\theta - \cos\theta)}{\cdot (ix + jy_\alpha + ky_\beta)} \right) \right] dm \right)$$

$$+ \left(\frac{F_{uv_{y\alpha}} M_{y\alpha}}{R^2} \int_{V_{y\alpha}} (Ry_\alpha (\sin\theta + \cos\theta)) \left[1 + \frac{3}{R} \left(\frac{j(\sin\theta + \cos\theta) - k(\sin\theta - \cos\theta)}{\cdot (ix + jy_\alpha + ky_\beta)} \right) \right] dm \right) \quad (1.870)$$

$$+ \left(\frac{F_{uv_{y\beta}} M_{y\beta}}{R^2} \int_{V_{y\beta}} (Ry_\beta (\sin\theta + \cos\theta)) \left[1 + \frac{3}{R} \left(\frac{j(\sin\theta + \cos\theta) - k(\sin\theta + \cos\theta)}{\cdot (ix + jy_\alpha + ky_\beta)} \right) \right] dm \right)$$

or also as

$$M_T = \frac{F_{uv_T} M}{R^2} \iiint_V dm (j^2 - k^2) \sin\theta \cos\theta \qquad (1.871)$$

or written vectorially as

$$M_T = \frac{F_{uv_T} M}{R^2} (C - A) \sin\theta \cos\theta = M_T = \frac{F_{uv_T} M}{R^2} (C - A)(R \cdot k)(R \times k) \qquad (1.872)$$

We now will transform M to the rest frame $OX_1Y_1Z_1$,in which the Earth describes its orbit about the Sun in the ecliptic plane OX_1Y_1 and the X_1-axis is chosen in the direction of the vernal equinox. Now the equatorial plane of the Earth, in the frame $OXYZ$, attached rigidly to the Earth, intersects the ecliptic along OX with an inclination θ, while $\angle(OX_1, OX) = \psi$. Draw OM orthogonal to OX and lying in the ecliptic plane. Then let

$$i_1 \text{ and } j_1 \text{ be unit vectors along } OX_1 \text{ and } OY_1 \qquad (1.873)$$

and

$$j' \text{ is a unit vector along } OM \qquad (1.874)$$

349

The given dispositions of the various vectors and the relations among them are exhibited in the diagram. We observe then

$$i_1 = OB = OA + AB = i \cos\psi - j' \sin\psi$$
$$j' = OF = OE + EF = i \cos\theta - k \sin\theta$$

(1.875)

Therefore

$$i_1 = i \cos\psi - j' \sin\psi \cos\theta + k \sin\psi \sin\theta$$

(1.876)

similarly, then

$$j_1 = OD = OC + CD = j \cos\psi - i \sin\psi = i \sin\psi + j \cos\theta \cos\psi - k \cos\psi \sin\theta$$

(1.877)

We assume that the Earth describes its orbit about the Sun in an elliptical orbit with an angular velocity ω. The position vector, R, of the Sun at a time t, after the vernal equinox, is given as

$$R = R\left(i_1 \cos\omega t + j_1 \sin\omega t\right)$$

(1.878)

Then substituting for i_1 and j_1 form equation (1.875) we find:

$$R = R\left[\begin{array}{l}\left(i \cos\psi - j' \sin\psi \cos\theta + k \sin\psi \sin\theta\right)\cos\omega t \\ +\left(i \sin\psi + j \cos\theta \cos\psi - k \cos\psi \sin\theta\right)\sin\omega t\end{array}\right]$$
$$= R\left[i \cos\left(\omega t - \psi\right) + j \cos\theta \sin\left(\omega t - \psi\right) - k \sin\theta \sin\left(\omega t - \psi\right)\right]$$

(1.879)

Then returning to equation (1.872), we find:

$$R \times k = \pm R \sin\theta \sin\left(\omega t - \psi\right)$$

(1.880)

and

$$R \times k = R\left[i \cos\theta \sin\left(\omega t - \psi\right) - j \cos\left(\omega t - \psi\right)\right]$$

(1.881)

We will then have

$$M_T = \frac{F_{uv_T} M}{R^2}(C-A)\sin\theta\sin(\omega t - \psi)\left[i\,\cos\theta\sin(\omega t - \psi) - j\,\cos(\omega t - \psi)\right] \quad (1.882)$$

Finally, as it is evident from the prior illustration,

$$\omega_x = \theta \qquad \text{and} \qquad \omega_y = \psi\sin\theta \quad\quad (1.883)$$

The equations for θ and ψ will follow from equation (1.859). We begin with

$$H_z = C\left(\phi + \omega_z\right) \quad\quad (1.884)$$

and then

$$C\frac{d}{dt}\left(\phi + \omega_z\right) = 0 \quad\quad (1.885)$$

We will then find

$$\left(\phi + \omega_z\right) = \Omega \quad\quad (1.886)$$

Where Ω denotes the angular velocity of the daily rotation of the Earth. We will then have for the two remaining equations of (1.859)

$$A\omega_x - A\omega_y\omega_z + C\omega_y\Omega = -\frac{f(t)(\beta)}{r^3}(C-A)\sin\theta\cos\theta\left[1 - \cos 2(\omega t - \psi)\right]$$

$$A\omega_y + A\omega_x\omega_z - C\omega_x\Omega = -\frac{f(t)(\beta)}{r^3}(C-A)\sin\theta\sin 2(\omega t - \psi) \quad\quad (1.887)$$

Then since the rotation of the frame $OXYZ$ is relatively slow, we can neglect the first two terms on the left-hand sides of equation (1.887) and then obtain:

$$\psi = -\frac{f(t)(\beta)}{r^3}\left(\frac{(C-A)}{C}\right)\cos\theta\left[1 - \cos 2(\omega t - \psi)\right] \quad\quad (1.888)$$

and

$$\theta = -\frac{f(t)(\beta)}{r^3}\left(\frac{(C-A)}{C}\right)\sin\theta\sin 2(\omega t - \psi) \qquad (1.889)$$

The mean rate of precession, $\langle\psi\rangle_{AV}$, is given by

$$\langle\psi\rangle_{AV} = -\frac{f(t)(\beta)}{r^3}\left(\frac{(C-A)}{C}\right)\cos\theta \qquad (1.890)$$

If the Earth is considered as an oblate spheroid with semiaxis a and $b(<a)$, then the moments of inertia C and A are:

$$C = \frac{2}{5}Ma^2 \qquad \text{and} \qquad A = \frac{1}{5}M\left(a^2 + b^2\right) \qquad (1.891)$$

Therefore

$$\frac{1}{C}(C-A) = \frac{1}{2a^2}\left(a^2 + b^2\right) = \frac{a+b}{2a}\left(1 - \frac{a}{b}\right) \qquad (1.892)$$

If the ellipticity

$$\varepsilon = (1 - b/a) \quad 1 \qquad (1.893)$$

then to the first order in ε,

$$(C-A)/C = \varepsilon \qquad (1.894)$$

then equation (1.890) gives

$$\langle\psi\rangle_{AV} = -\varepsilon\frac{f(t)(\beta)}{r^3}\cos\theta \qquad (1.895)$$

Finally, then, the angular velocity of the daily rotation of the Earth is

$$\Omega = 2\pi \times 366\frac{1}{4} \quad \text{radians per year} \qquad (1.896)$$

and the inclination of the Earth's rotation axis to the ecliptic is

$$\theta = 23 \cdot 45 = \cos^{-1} 0 \cdot 91741 \qquad (1.897)$$

(d) The precession due to the presence of the Sun

Because the rotation and its magnitude is due to the mass of the Earth and the gravitational field generated by the presence of the Sun, we will then have

$$\frac{\left(F_{uv_\alpha} \oplus F_{uv_\beta} \right) M}{R^3} = \left(\frac{2\pi}{T} \right)^2 \qquad (1.898)$$

then by Kepler's third law, $T \left(= 365\frac{1}{4} \right)$ days, is the period of revolution of the Earth about the Sun. Inserting these values (1.896-1.898) in equation (1.896) and setting

$$\varepsilon = 1/305 \cdot 6 \qquad (1.899)$$

we will have then

$$\langle \psi \rangle_{AV:} = -\frac{3\left(4\pi^2\right) \times 0 \cdot 91741}{2\left(2\pi \times 366\frac{1}{4} \right) \times 305 \cdot 6} = -7 \cdot 725 \times 10^{-5} \qquad (1.900)$$

or

$$\langle \psi \rangle_{AV;} = -15 \cdot 94'' \quad \text{per y ear} \qquad (1.901)$$

If we choose

$$\varepsilon = 1/230 \qquad (1.902)$$

as given for a homogenous Earth (Chapter 20, section 109) in place of (63) we will have then

$$\langle \psi \rangle_{AV;} = -15 \cdot 94'' \frac{305 \cdot 6}{230} = -21 \cdot 18'' \tag{1.903}$$

(e) The lunar contribution to the precession and the lunisolar precession

If we dentify s with the Moon, then we will have

$$\frac{\left(F_{uv_\alpha} \oplus F_{uv_\beta}\right)M - \left(F_{uv_\alpha}\right)M_M}{R_M^3} \frac{M_M}{M + M_M} = \left(\frac{2\pi}{T_M}\right)^2 \frac{1}{82 \cdot 5} \tag{1.904}$$

where

$$T_M = \frac{27 \cdot 32}{365 \frac{1}{4}} \text{year} \tag{1.905}$$

is the period of revolution of the Moon about the Earth and

$$82 \cdot 5 = 1 + M / M_M \tag{1.906}$$

Inserting the values (58-59), (61), (66), and (67) in equation (58) we obtain then

$$\langle \psi \rangle_{AV:Moon} = -\frac{3\left(4\pi^2\right) \times 0 \cdot 91741}{2\left(2\pi \times 366\frac{1}{4}\right)82 \cdot 5}\left(\frac{365\frac{1}{4}}{27 \cdot 32}\right)^2 \frac{1}{305 \cdot 6} = -1 \cdot 674 \times 10^{-4} \tag{1.907}$$

or

$$\langle \psi \rangle_{AV:Moon} = -34 \cdot 5'' \text{ per year} \tag{1.908}$$

Equations (62) and (69) together give for the total lunisolar precession the value

$$\langle \psi \rangle_{AV:Total} = -50 \cdot 4'' \text{ per year} \tag{1.909}$$

in close agreement with the observed value of $-50''$ per year.

At the rate of precession (71), the time required for a complete revolution of the vernal equinox about the Sun is $25,700$ years.

The ratio of the tidal effects due to the presence of the Sun and the Moon is given as

$$\frac{\langle\psi\rangle_{AV:Moon}}{\langle\psi\rangle_{AV:}} = \frac{34\cdot5''}{15\cdot93''} = 2\cdot164 \tag{1.910}$$

Finally if we had chosen for ε the value $\dfrac{1}{230}$, then for a homogenous Earth we would have obtained

$$\langle\psi\rangle_{AV:Moon} = -45\cdot83'' \text{ per year} \tag{1.911}$$

and

$$\langle\psi\rangle_{AV:Total} = -67\cdot0'' \text{ per year} \tag{1.912}$$

Moment of momentum, moment of inertia, and circulation

In Lemmas I and II, we brought forth the notions of moment of momentum and moment of inertia in the context of a rigid body with an angular velocity Ω about a fixed axis passing through its center of mass.

Lemma I

If $APEp$ represents the Earth as uniformly dense, marked with the center C, the poles P, p, and the equator AE; and if about the center C, with the given radius CP, the sphere $Pape$ is described, and QR to denote the plane on which a right line, drawn from the center of the Sun to the center of the Earth, stands at given right angles; further then the separated particles of the whole exterior Earth $PapApepE$, without the height of the that sphere, endeavor to recede towards this side and that side from the plane QR, every particle by a force proportional to its distance from that plane. The whole and efficacy of all the particles that are situated in AE, the circle of the equator, and disposed uniformly without the globe, encompassing the same after the manner of a giver ring, to wheel the Earth about its center, is to the whole force due to the presence of the Sun according to the laws of the $G.R.F.F.$ as many particles in that point A of the equator which is at the greatest distance from the plane QR. The same force is to wheel the Earth about its center with a circular rotational motion. This circular motion is to be performed about an axis lying in the common section of the equator and the plane QR.

The Lemma states that the velocity v of an element of mass, dm at a point r, in a rigid body rotating with an angular velocity Ω, is given as

$$v = \Omega \times r \tag{1.913}$$

while its momentum of moment or angular momentum, is given by

$$dH = dmr \times \left(\Omega \times r\right) \tag{1.914}$$

The angular momentum, H, of the whole body is then given as

$$H = \int_V dmr \times \left(\Omega \times r\right) \tag{1.915}$$

here the integral is extended over the volume V occupied by the whole body.

The definitions (1) and (2) are explained in Lemma I in the apparent context of a plane circular disc of radius a rotating with an angular velocity Ω in a direction that is normal to the disc. An element of mass, $dM(r)$, at a point P on the disc at r has a velocity ω of magnitude

$$\omega = \Omega r \sin \theta \tag{1.916}$$

normal to the plane $CPQ\Omega$. Then, its moment of momentum,

$$dH(r) = dM(r)\Omega r^2 \sin\theta \qquad (1.917)$$

which lies in the plane $CPQ\Omega$ and is parallel to the direction Ω. The given component of this moment of momentum normal to the direction CQ, is

$$dH_r(r; \perp^r CQ) = dM(r)\left[\Omega r^2 \sin\theta\cos\theta\right] \qquad (1.918)$$

If the same mass $dM(r)$ is placed at the circumference of the disc, its moment of momentum will then be

$$dH_a(a; \perp^r CQ) = dM(r)\left[\Omega a^2 \sin\theta\cos\theta\right] \qquad (1.919)$$

The ratio of the expressions (6) and (7) is independent of θ and is then given by

$$dH_r(r; \perp^r CQ) : dH_a(a; \perp^r CQ) = dM(r)\Omega r^2 : dM(r)\Omega a^2 \qquad (1.920)$$

Then the ratio of the moment of momentum of the entire disc to the moment of momentum of all the mass distributed over the circumference is given as

$$\int_0^a dH_r(r; \perp^r CQ) : dH_a(a; \perp^r CQ) = \Omega\int_0^a dM(r)r^2 : Ma^2\Omega \qquad (1.921)$$

where M denotes the mass of the disc. If the disc is of uniform density, the mass, $M(r)$, interior to r is

$$M(r) = Mr^2 / a^2 \qquad (1.922)$$

and

$$\Omega\int_0^a dM(r)r^2 = \frac{\Omega a^2}{M}\int_0^a dM(r)M(r) = \frac{1}{2}M\Omega a^2 \qquad (1.923)$$

From equation (9) it follows that

$$\int_0^a dH_r\left(r;\perp^r CQ\right):H_a\left(a;\perp^r CQ\right)=1:2 \qquad (1.924)$$

We conclude according to Newton then:

And because the action of those particles is exerted in the direction of lines perpendicular and receding from the plane QR, and equally from each given side of this proposed plane, they will then wheel about the circumference of the circle of the equator, together with the adherent body of the Earth, round axis which itself lies in the plane QR, as in that of the equator.

The moment of inertia of a uniform circular disc about its normal is, noticed by Newton, as

$$I = \frac{1}{2}Ma^2 \qquad (1.925)$$

He also proved the invariance of the ratio (9) to the angle of inclination θ of the reference direction CQ: which is essential to the identification of $\frac{1}{2}Ma^2$ as the force that wheel's the disc.

Lemma II

The total force or power of all the particles situated everywhere about the sphere in relation to the force that turns the Earth about its axis is to the same force of the number of particles that are uniformly disposed round the whole circumference of the equator AE in the fashion of a ring, all of which is generated by the $G.R.F.F.$.

Here in this Lemma, we evaluate the angular momentum of at rotating homogenous spherical body of mass M and radius a about the axis of rotation along the $Z-$axis.

Then by the definition of this quantity in question we will have

$$M_z = 4\pi \int_0^a \int_0^\pi \rho \left[r \times \left(\Omega \times r \right) \right]_z r^2 \sin\theta \, d\theta \, dr \tag{1.926}$$

Then since

$$\left[r \times \left(\Omega \times r \right) \right] = \left[\Omega r^2 - r^2 \left(r \cdot \Omega \right) \right] = \Omega \left(r^2 - z^2 \right) \tag{1.927}$$

then we will have

$$M_z = 4\pi \Omega \rho \int_0^a \int_0^\pi r^4 \sin\theta \cos\theta^2 \, d\theta \, dr = \frac{8\pi}{15} \Omega a^5 \rho \tag{1.928}$$

or

$$M_z \left(sphere \right) = \frac{2}{5} M \Omega a^2 : \left(M = \frac{4}{3} \pi \rho a^3 \right) \tag{1.929}$$

Though if the mass M is distributed uniformly over the circumference of the equator, as a given spherical shell, its angular momentum will then be given as

$$M_z \text{(circumferential ring)} = M \Omega a^2 \tag{1.930}$$

Then

$$M_z \left(sphere \right) : M_z \text{(circumferential ring)} = 2 : 5 \tag{1.931}$$

Lemma *III*

The motion of the Earth about the given axis, arising from the motion of all the collected particles constituent of the Earth according the laws of the *G.R.F.F.*, will be to the motion of the previously mentioned ring about the same axis in a ratio compounded of the ratio of the matter in the Earth to the matter in the ring as determined for both directly by the magnitude of the *G.R.F.F.*. The ratio of three squares of the quadrantal arc of any circle are to two squares of its diameter, that is to say in the ratio of the matter to the matter, and of the number 925275 to the number $1,000,000$.

Here the notion of circulation is more readily proposed. By the notion of 'motion' we here mean in the context as given by Newton, circulation, also as the line integral of the velocity is along a closed curve. Here the circulation along a circle of radius a normal to the axis of rotation, is (by equation 4)

$$C_z(r) = 2\pi\Omega r \sin\theta \qquad (1.932)$$

Therefore, the circulation of the motions in the whole sphere is

$$C_z = 2\pi\Omega \int_0^a \int_0^\pi r^3 \sin^2\theta \, d\theta \, dr = \frac{1}{4}\pi^2 a^4 \Omega \qquad (1.933)$$

or

$$C_z = \frac{3\pi}{16} Va\Omega \qquad (1.934)$$

where V denotes the volume $\left(= \frac{4}{3}\pi a^3\right)$ of the sphere.

At the same time, the circulation, ΔC, of a thin circular ring of radius a and thickness

Tidal Forces

Proposition LLVII. Theorem I

The flux and reflux of the sea arise from the presence of both the Sun and the Moon

Waters of the sea rise and fall twice for every day, as both due to lunar and solar contributions. The radiation from th Sun on the Earth causes the waters to fall be pushed sideways and thus recede and reduces the bulge formed due to the presence of the Moon. The presence of the Moon shields the radiation from the Sun, and thus with no apparent extra force on the surface of the oceans, and the radiation emanenating from the Earth, pushes the waters outwards, thus increasing the apparent volume of the water at that location of the surface of the earth, producing a bulge proportional to how much radiation is shielded.

The greatest height of the waters in the ocean follow the presence of the luminaries to the meridian of the given place by an interval less then six hours, all of which is in relation to the distance between the Earth and the Moon. The shores as the flood falls out about the second, third, and fourth hour, unless where the given motion propagated from the ocean, by the shallowness of the channels, through which it passes to some particular places, is retarded in the later hours, all proportional to the distance between the Earth and the Moon. The given hours from the position of the Moon to the meridian of the place, both above as well as below the horizon, the 24th parts of the time which the Moon by its diurnal motion, comes about the meridian of the place which it left the day before. The force of the Sun reduces the bulge of the oceans in proportion to the magnitude of the radiation, and the presence of the Moon shields that radiation, and thus produces an apparent bulge of the oceans. The position of the Moon causes the sea to rise, in terms of its position to the meridian of that place. The closer the Moon is, the greater the bulge and height at which the sea rises.

As the presence of the Moon raises the waters, in direct proportion to its distance from the Earth. From the syzygies and quadratures, will be the greatest tide due to the presence of the Moon. Then if the Moon passes syzygies to the quadratures, the greatest height of the waters will precede due to the presence of the Moon, and by the greatest interval, a little after the octants of the Moon, and by like intervals, the greatest tide will follow due to the presence of the Moon as its passes from the quadratures to the syzygies.

The effects of the presence of the Moon, depend on its distance from the Earth. When they are less distant, the effects are greater and when they are greater distance the effects are greater. That is the cube of their apparent diameter. The Moon, when in the perigee raisese greater tides then when it is at its apogee. The effects of both the Sun and Moon depend on their distance or declination from the equator; the if the body of the Moon was placed at the pole, it would constantly affect all parts of the waters. Then as both bodies decline from the equator towards either pole they will by degrees, lose their force and will excite lesser tides in the solstitial than in the equinoctial syzygies. In the solstitial quadratures the Moon will raise greater tides. All in proportion to its distance from Earth.

19

On comets

Introduction

Here we focus on the curve of a comet in terms of the parabolic order, through some given points, in terms of some newly derived solutions.

Lemma IV and Proposition LLVIII

Lemma IV

The comets of more remote then the moon, and are in the given regions of the planets.

Let *YQA, YQB, YQC* be three given longitudes of the comets about some point of trajectory, and at *YQC* as the last point of longitude. We can draw the right line, *ABC*, whose parts, *AB, BC*, intercepted between the right lines *QA* and *QB* and *QC*, are one to another. Form *AC* to *G*, in which *AG* is to *AB* and thus join *QG*. If the comet were to move in a right line, and the Earth were to remain still, or moved in straight line, then the angle *YQG* would be the determined longitude of the comet. Then the angle *FQG*, as the difference in longitude, proceeds from the given inequality of the motions of the comets and the Earth; and the Earth and comet move contrary ways, this given angle is then added to the angle *YQG*, and accelerates the motion of the comet according to the *G.R.F.F.* Though if the comet moves with the Earth, is then subtracted, it then retards the motion of the comet. This angle, proceeds directly from the motion of the Earth, as the parallax of the comet, then it there be neglected some increment or decrement arising from the unequal motion of the comet in its orbit. From here we can determine the distance of the comet. Let *S*, represent the Sun, *acT* the orbit, a the Earth'position in the first observation, c the place of the Earth in the third observation, T the place of the Earth in the last observation, and *TY* a right line drawn to the beginning of the observations. Then set of the angle *YTV*, equal to the angle *YQF*, that is, equal to the longitude of the comet at the time when the Earth is in *T*. Then join *ac*, and produce it to *g*, where ag may be to ac as *AG* is to *AC*; and g as the place at which the Earth would arrive at the last observation, if it had continued to move in right line ac. If we draw *gY*, parallel to *TY*, and make the angle *YgV*, equal to the angle *YQG*, this angle *YgV* will be equal to the longitude of the comet seen from the place *g*, and the angle *TVg* will be the parallax which arises from the Earth being moved from the place *g* intot the place *T*; and then *V* will be the place of the comet in the plane of the ecliptic. All of this is determined by the value of the *G.R.F.F.* and the intial velocity of the comet before entering the field of the *G.R.F.F.*

This celestial body receding to remote parts, diminishes as fourth power of the distance; namely as the square, in terms of the increase of the distance from the Sun, and another square, on account of the decrease of the apparent diameter, all of which is determined by the magnitude of the *G.R.F.F.*

Proposition LLVIII. Theorem I

Given that the comets move in some conic sections with their foci in the center of the Sun, and by radii drawn to the Sun, describes areas proportional to the times.

Cor. I. Hence if the comets revolve in orbits returning to themselves, the orbits will then be ellipses; and their periodic times will be to the periodic times of the planets as the 3/2 power of their principle axes. Therefore, the comets, as they move far from the planets, describe orbits with greater axes, will need a longer time to finish their revolutions as determined by their initial velocity and the magnitude of the *G.R.F.F.* If he axis of the comet's orbit were four times greater then the axis of the orbit of Saturn, the time of revolution of the comet would be the time of the revolution of Saturn, 240 years.

Cor. II. Their orbits are near to parabolas, as determined by the *G.R.F.F.*

Cor. III. The velocity of every comet will then be added to the velocity of every planet, as supposed to be revolved at the same distance in a circle about the Sun, nearly as the square root of double the distance of the planet from the center of the Sun to the distance of the comet from the center of the Sun, as determined by the *G.R.F.F.* Given the radius of the great orbit, or the greatest semidiameter of the ellipse which the Earth describes, to consist then of 100000000 parts; then the Earth describes by its mean diurnal motion will describe 1720212 of those parts, and 71675(1/2) by its hourly motion. The comet then, as the same mean distance, of the Earth from the Sun, with a given velocity in which it would by its diurnal motin describe 2432747 parts, and 101364(1/2) parts by its hourly motion. At greater or less distances, both the diurnal and hourly motion will be to this diurnal and hourly motion inversely as the square root of the distances, as determined directly by the magnitude of the *G.R.F.F.*

Cor. IV. If the latus rectum of the parabola is four times the radius of the great orbit, and the square of that radius is to consist of 100000000 parts, the area in which the comet will daily describe by a radius drawn to the Sun, according to the *G.R.F.F.*, will be 1216373(1/2) parts, and the hourly area will be 50682(1/2) parts. Though if the latus rectum is greater or less in a given ration, the diurnal and hourly area will be less or greater inversely as the square root of that ratio, as determined by the magnitude of the *G.R.F.F.*

Lemma V

We seek to find the curved line of the parabolic kind in which shall pass through any given number of points.

We must find a polynomial curve that will pass through the extremities, *A, B, C, D, E* and *F* of the ordinates, *AH, BI, CK, DL, ME, NF*; and find the value of the ordinate *RS* at the intermediate point, *R* on the given curve.

Lemma VI

Certain observed places of a given comet, we seek to find the place of the same at any intermediate given time, according the values of the G.R.F.F.

Let *HI, IK, KL, LM* represent the times between observations; *HA, IB, KC, LD, ME*, five observed longitudes of the comet inaccordance with the values of the G.R.F.F; and *HS* the given time between the first observation and the longitude needed. Then given a regular curve *ABCDE* is to be drawn through the points *A, B, C, D, E*, and the ordinate *RS* is found to be the longitude needed.

If the differences are small, 4 to 5 degrees, three or four observations will be needed; but if they are greater, 10 to 20 degrees, all observations should be used

REFERENCES:

Joos, Georg, Freeman, Ira. Theoretical Physics. New York: Hafner Publishing Company, 1958

Black, Ethan D. A First Course in Geometric Topology and Differential Geometry. Boston: Birkhauser, 1997

Fock, Vladimir Aleksondrich. The Theory of Space, Time, and Gravitation. New York: Pergamon Press

Lorenz, H. A. Minkowski H, Eistein A., Newton Isaac, and Galileo. On the Shoulders of Giants: Dialogues Concerning Two New Sciences, Principia, and The Principle of Relativity. London: Running Press, 2002

Jackson, John David, Classical Electrodynamics. New York: John Wiley & Sons Inc., 1962

Lucas J. R. and Hodgson P. E. Spacetime and Electromagetism. Oxford: Clarendon Press, 1990

Kriele, Marcus. Spacetime: Foundations of General Relativity and Differential Geometry.

Heinbackel, H. John. www.math.odu.edu/~jhh/counter2.html

www.scienceworld.wolfram.com/physics/topics/pendula.html

www.scienceworld.wolfram.com/physics/topics/fluidmechanics.html

www.scienceworld.wolfram.com/physics/topics/dynamicsandkinematics.html

www.scienceworld.wolfram.com/physics/topics/specialrelativity.html

www.pancake.uchicago.edu/~carroll/notes/html

Moller, Christian. The Theory of Relativity. Dehli, Bombay, Calcutta, Madras: Oxford University Press, 1972

Logunov, Anatolii Alekseevick. <u>Lectures in Relativity and Gravitation: A Modern Look</u>. Moscow: Nauka Publishersm 1990. Oxford, New York, Beijing, Frankfurt, Sao Paulo, Sydney, Tokyo, Toronto: Pergamon Press.

Tullio Levi-Civita. <u>The Absolute Differential Calculus</u>. New York: Dover Publishing Inc., 1997.

Ray, M. <u>Theory of Relativity: Special and General</u>. Delhi, New Delhi, Jullundur, Lucknow, Bombay, Calcutta, Madras, Hyderabad: S. Chan & Co.

Schwinger, Julian; Lester, DeRaad Jr.; Milton A. Kimball; Tsai Wu-Yang. <u>Classical Electrodynamics</u>. New Jersey, London, Singapore, Beiging, Shanghai, Hong Kong, Taipei, Chennai: World Scientific, 2004.

D. S. Jones. <u>The Theory of Electrodynamics</u>. New York: The Macmillan Company, 1964.

Jones, Oliver Davis. <u>Analytical Mechanics for Relativity and Quantum Mechanics</u>. Oxford University Press, Oxford, 2005.

S. Chandrasekhar. <u>Netwon's Principa for the Common Reader</u>. Trinity College, Cambridge.

www.ingramcontent.com/pod-product-compliance
Lightning Source LLC
Chambersburg PA
CBHW081715220526
45468CB00008B/1848